"Let's Hit the Trail Together"

I Married a Logger

Life in Michigan's Tall Timber by

Julie Anderson

With Illustrations by the Author

Printed in the United States of America
Avery Color Studios

Reprinted by Avery Color Studios
Au Train, Michigan 49806

March 1988

Library of Congress Catalog Card Number: 51-11829
ISBN# -932212-54-9

Cover art by Keith Avery

This story is dedicated to
the logger I married
and our three jacks,
Chris, Hans, and Hayward.
J. A.

Photo by Ken Bauer

Mrs. Anderson, now 76 years old, is currently a resident of the Luther Home in Marinette, Wisconsin. Her Howard, after leaving the logging camps, sold logging equipment in Wakefield, Michigan, then worked as a machinest in Green Bay, Wisconsin until his retirement in 1975. His health slowly declined however, and he died in February 1986.

Julie's "three jacks" are scattered with Chris living in the Northwest, Hans, an electrical engineer, lives in Bloomington, Minnesota. The youngest, Hayward, is a partner in a pharmaceutical business in Marinette, Wisconsin. There are eleven grandchildren.

Acknowledgments

During the process of preparing this manuscript and the sketches, I have become indebted to a number of people. I am most grateful to Howard my husband and our three young sons—Chris, Hans, and Hayward. They have all of them tolerantly and understandingly coped with a preoccupied and daydreaming wife and mother. To them I owe years of love and indulgence—an obligation I will find pleasure in repaying.

Without the recollections of Mrs. Wescott, Mrs. Louise Goodenow, Mr. Fred Gregory, Mr. August Bye, Mr. and Mrs. Fred Brandt, "Doc" Masterson, Mr. Bill Weston, Mr. Tom Smith, Mr. Jim Gough, Mr. George Holmes, Mrs. Viola Haug, Mr. Victor Lepisto, Mrs. Isabel Anderson, and my husband Howard, my story would be decidedly less colorful. For their contributions I am especially grateful.

To the lonely transient jacks of Tula Camp, whose combined stories grew into "Booze, Broads, and Brawls," and who found my attentive listening thanks enough for their tales, I wish to express my written appreciation.

I particularly wish to express my appreciation to Professor Ralph Alan McCanse of Wisconsin Extension Service. His effective criticism and friendly suggestions encouraged me and provided the fundamental basis for improving my story during months of rewriting.

To Miss H. Shirley, Mr. H. Trezise, Mrs. A. Bodah, Mrs. W. Carow, and my sister Mrs. H. Williamson, I am indebted for valuable professional services rendered without remuneration. The Wakefifield librarian, Mrs. Margret Taylor, and Mr. Edwin Johnson of the *Ironwood Daily Globe* staff have both been very helpful.

I have found of great value the advice and encouragement that came to me from Mr. William Atwood, one of my former instructors and, through the years, a true and helpful friend. He at one time wrote me, "Now, getting a book published is one of those great lotteries in life. Sometimes you win, and sometimes you lose; and it is very difficult to tell in advance what will be popular and what will not, or what a publisher will take. So, if you do not find an immediate market, do not be discouraged, but persist."

Because of the interest, encouragement, and contributions of the above-named persons and many other friends, I have indeed persisted. The experience has been pleasant and exciting. The result is this following record of happy invigorating years of being married to my logger Howard.

Contents

PART FOUR

PART FIVE

List of Illustrations

In the Northland

Oh the sun shines bright and the world is white
And the wind blows cold from the north;
It is time to go where the ice and snow
Hold a thrill if we will go forth;
Where the ax swings free, and the proud old tree
Comes to earth with a mighty roar.
Where a hearty song speeds the work along,
And a man is a man once more.

Oh the north wind blows, and the drifting snow
Brings a thrill and a zest to the game;
There is work to do, there is danger too,
But to fear puts a man to shame;
For the sun shines bright and the world is white,
And the forest is calling me;
It is time to go to the land of snow,
Where the heart of a man is free, is free;
Where the heart of a man is free.

CLAY SMITH

Sunshine in My Wisconsin

Without the lumbermen's realizing it, the years of unrestricted logging of white pine had depleted the Michigan and Wisconsin forests, until at the turn of the century there remained little of the vast wilderness which the lumbermen had hoped could not be logged in a century. Only stumps remained where once virgin pine had stood. The fertile lands which had produced enormous pines then lured pioneers of many nationalities into farming. My Grandfather Benson, an English immigrant, came to Wisconsin when pine was becoming scarce and the broad, rolling fields from which the timber crop had been harvested lay ready to begin the cycle of small crop production. My Grandfather Clark was one of many lumbermen who, after logging their land, realized that the timbered areas were fast decreasing in size and, rather than leave Wisconsin, proceeded to remove the stumps from their land and prepare it for tilling and planting. Other lumbermen turned their eyes to the beckoning giants on the West Coast and migrated to that new timberland. Still other lumbermen, reluctant to span the prairie in search of timber, sent their cruisers north to the Upper Peninsula of Michigan, where they found that the pine was less dense, but, to even the measure, there were vast growths of hardwood and hemlock.

As loggers gradually progressed from the eastern part of the peninsula westward, there began a new era in Michigan logging during which loggers felled the remaining pine and adventured into cutting hardwoods. River transportation of logs was coming to an end because most pine stands along the waterways were cut. The inaccessibleness of the decreasing stands of pine forced the loggers to build railroads and use locomotives as a means of getting logs out of the forests. During the years when Michigan lumbermen turned to the Upper Peninsula for their pine logs and began building railroads to the remote tracts of timber, Isabel and Louie Anderson, my husband's parents, came to Ewen, Michigan, to live. Also at that time my mother and father left Wisconsin to build their home on a Kincaid claim near Kimball, Nebraska, where my sister Bertha and I were born.

The same elements harassed my folks in Nebraska that caused timber damage and property loss for Louie Anderson and other loggers in Michigan. Fire, which swept through the dry forests on hot autumn days, devastating millions of dollars' worth of white pine and leveling logging camps, lumber mills, and lumber yards, blazed up in the tinder-dry sagebrush and rolled across the prairie to consume my father's wheat fields, stampede his horses, and burn his neighbors' homes and barns. Vaguely I remember (partly from hearing my father tell the story many times) crouching near my mother in the root cellar while Dad plowed the neglected firebreak. Yelling wildly and lashing his span of terrified horses with a seldom-used rawhide whip, he plowed until he was exhausted; but his firebreak checked the rolling flames.

Wet, heavy snow became so deep in Michigan that men and horses could not work in the woods. Fine powdery snow, drifted by blizzard winds howling across the prairie land, isolated us from our well, our barns, our chicken coops, and from neighboring claims.

Winds blew high through the pine stands and caused inestimable timber damage. Winds swept low across Nebraska, scattering the tumbleweeds, leveling wheat fields, and lifting buildings from their foundations.

Inexorable changing seasons, with their blazing suns, blizzards, and tornadoes, gave my father the color of a hazelnut, made him tough as an oak and tolerant of my tempestuousness. Scanning the flat horizon for the blaze of prairie fires, the curl of a twister wind, or the last rays of a flamboyant sunset, gave his eyes keen sight and his heart patience. Life on a bleak western prairie claim gave my mother an intense love for trees, blue lakes, and neighbors.

The year the prairie fire threatened our claim my father pulled stakes. A new frontier in the unclaimed areas of Canada lured him to that wilderness to raise wheat. When I was five years old we left our golden, snake-infested, windswept wheat fields and jolted along in a cart behind a yoke of oxen on our journey to a log cabin in Alberta's fertile but unbroken land.

My father, a husky, barrel-chested farm product, planned to clear his Canadian claim—a small piece each year—until he had many acres of fertile wheatland. His dreams might have been realized if my mother could have tolerated the howling of the timber wolves; the depressing minor song of the wind in the pines; arduous washing of clothes with no soap, on occasions when the supplies were gone; living on moose meat, tea, and biscuits; and the complete isolation from schools and neighbors. In addition to my mother's unhappiness, a toothache which my father suffered all during the second winter brought about our departure from the Canadian wilderness. That spring we returned to Wisconsin, my parents' native state, where my father established himself in the dry goods business in Shell Lake, a small resort town which had once been a booming mill town before the area around it was stripped of its pine by lumbermen and by fires.

I grew up in the pleasant atmosphere of modest prosperity, parental love, and Wisconsin sunshine. In summer my sister and I swam in the sandy-bottomed lake, and in winter we skated on its glistening ice. Because I enjoyed such sports, I stayed slender. (Or, perhaps, because I was naturally agile and slender, I liked to swim and skate.) A deep hollow and

A Dollar a Day and His Chuck

the paths along a tiny stream that curved around the wooded hills behind the schoolhouse were my favorite haunts when I was not swimming, skating, drawing, or practicing music lessons.

As my father prospered, he bought a Chevrolet touring car that took us over narrow, sandy roads and brought us home safely from many family expeditions into Wisconsin's lake and stream regions. At the unromantic hour of five-thirty Sunday mornings, Bertha and I, still half-asleep, would climb out of bed and struggle into hiking bloomers and blouses, long black stockings, and sturdy shoes. While Mother packed a picnic basket with chicken, rolls, salad, and gallons of lemonade, Bertha and I gobbled breakfast and climbed into the car. Beside a trout stream or placid lake, where the black bass rose to Dad's exotically colored flies, Mother unpacked our lunch, and we spent the day. When the sun sank low we would climb back in the car, snuggle under the smelly buffalo robe, jolt over winding roads, and arrive home feeling tired but serenely happy.

As we girls grew older there were fewer family picnics. Instead Bertha and I canoed down the fast waters of the Namekogan River, the lazy Yellow River, or the Totogatic River, which Dad called the "Tote-your-own-damn-canoe River" because in dry weather it was too shallow in many places to float our canoe. Preferring not to portage, we paddled only once down the Totogatic River. Yellow River's course is through Rice Lake, a small marshy lake almost entirely covered with rice beds. As we paddled through these rice beds in the fall, we watched the Indians thresh the rice kernels into their canoes with narrow paddles, just as they had done for ages past.

Along the shore we saw their roughly built summer camps beneath the tall jack pines. Strings of fish hung on poles over smoky fires tended by the black-haired, dark-eyed Indian children. We called greetings to them as we floated by, but they were much too shy ever to answer. (I don't believe we could have understood them if they had!)

As most girls do, I can remember fervently wishing that I were beautiful. I admired my mother's black, shining hair and hated my slightly curly, light-brown locks. Secretly I tried everything in the drugstore to bleach my freckles. (I still have them today and still hate them.) My only happiness was that I had inherited Mother's English violet-blue eyes and jet-black eyebrows and lashes, and I didn't get lumpy with fat. It is a pity that I didn't inherit my father's Scotch qualities; instead, I am an incurable spendthrift.

Mother and Dad expected us to be obedient, kind, pleasant, polite, and alert. They encouraged us to make the most of our talents. They realized that things practical for some people are not at all practical for others. They never appeared to be discouraged when I was not so studious as Bertha; they would take pride, rather, in my devotion to beautiful things. In my home town I was merely Bertha's sister, who could sing and play the piano, who could draw prize-winning posters for the liquor prohibitionists, and who wasn't afraid of snakes. I recall that I was pathetically poor at spelling—words like *dining* and *coming* being the most difficult for some uncertain reason. Bertha, who was a brilliant student and generous in her defense of my inefficiencies, said it was not my fault if, every time the teachers chose to teach how to add *-ing* to words like *come* and *dine,* I happened inopportunely to be sick with the measles, whooping cough, or chicken pox. Loyal Bertha!

These numerous children's diseases weakened my eyes, and the result was my having to wear glasses—not the coquettish horn-rimmed kind the girls wear today in an attempt to look pretty, but terrible black-rimmed ones with round lenses, making me as attractive as an owl. I was always taking them off and leaving them on other people's pianos for weeks at a time, or putting them in my coat pocket and promptly sitting on them. Never have I reconciled myself to wearing the pesky things. When I at last persuaded my mother to let me enroll at the Chicago Academy of Fine Arts, the first thing I did was to buy a pair of oxford glasses with money I earned waiting on tables at the YWCA where I lived. Knowing that I could whisk them off easily at a moment's notice (and neither lose them nor break them in doing so), gave me confidence which had been smothered behind glasses all during my high-school days.

I chose illustration as my first love in the field of art—though I never expected to illustrate life in a logging camp from personal experience. My dreams were of becoming a successful fashion illustrator and of some day seeing Paris.

Suddenly the year came when I finished my training at the academy; with it came the memorable stock market crash. Overnight I was catapulted into the nightmare of failure, because I knew my chances of getting employment in Chicago were gone, along with the fortunes of many men.

The depression years, which changed the courses of many lives, forced me to abandon my hopes of seeing Paris through the discriminating eyes of a fashion illustrator. I enrolled at the Milwaukee State Teachers College as an escape from empty, idle days. To finance this adventure, I worked for my room and board. The stars that shone through the window of my attic room seemed unfriendly and vastly dimmer than the stars I had watched compete in brilliance with Buckingham Fountain and the lights on Michigan Avenue. Some weeks all I had in my purse were a few cents and a streetcar pass, nearly worn out from having been lent so many times to acquaintances even more unfortunate than I.

Shortly before I received my degree in art education, I signed a contract to teach art in Spooner, Wisconsin. (My uncle was on the school board, if that's of any meaning to the more skeptical of my readers.)

Not until teaching positions became more plentiful did I look for a better paying one. My desire for a larger salary led me to Michigan's rich lumbering and mining district in the Upper Peninsula to teach. Assured by the superintendent that accepting a position in Wakefield would be an advancement, I agreed to sign a contract which would make me a supervisor of art.

Compared to Chicago and Milwaukee—or even Spooner—Wakefield seemed, at first, like a jumping-off place. When I told my friends in Spooner where I expected to teach, they looked amazed and said they had never heard of the town. To reassure myself that there really was such a place, I found its location on a map and discovered it was just at the edge of Wisconsin, only a few miles from the then-notorious Hurley. I had heard enough horrible tales concerning what went on in Hurley during my high-school days to make the expression "Hayward, Hurley, and hell" quite meaningful. Since I found Wakefield to be just beyond Hurley, I felt that Wakefield must be quite a spot, too. I am ashamed of my skepticism now, but before I signed the contract I drove over to see what was beyond Hurley in Michigan.

I found this region a rough country rich in natural and man-made beauty with mine shafts rising above the southern horizon and silhouetted boldly against the clear blue sky. Great piles of red-purple ore loom up on all sides here, even now, until the shipping season depletes them. I found mile-long lagging piles gleaming like silver in the bright sunlight. Ore-red roads in this region, thoroughfares of helmeted miners and trucks piled high with hardwood and hemlock logs, curve through neat mining towns of identical white houses. To the north, the knobby foothills of the Porcupine Mountains still form a jagged skyline. These heavily timbered hills jut above ore pits and mine shafts and drop roughly down to the white sands, jagged rocks, and scraggly pines along Lake Superior's south shore.

The natural beauty of these hills and the Porcupine Mountains is enhanced by ancient and romantic Indian legends. Ojibways knew this lofty range rising from "Gitche Gumee"—the shining, big sea water made famous by Longfellow—as "Kaug Wudjoo," the "crouching porcupine." As the Indians paddled their birchbark canoes along Gitche Gumee's shores, the mountains looked to them like crouching porcupines. Years later, after explorers and *voyageurs,* fur traders and prospectors had passed this way to camp at the mouths of the Carp, Presque Isle, and Black Rivers, the name was shortened to "Kaug" and then translated into English—"the Porcupines." To me, it is synonymous with north woods.

The vine-covered brick schools and community buildings of the mining towns impressed me. I actually thought that what turned out to be Wakefield's high school could be nothing less than a mining college. A tour of the buildings, one look into the cupboards filled with art supplies, and a glance out of the schoolroom window toward Sunday Lake mine shaft convinced me that I could enjoy teaching in this colorful place beyond Hurley. Without further hesitation, I signed the contract and sealed my fate by renting a room from Isabel Anderson—who was to be my mother-in-law.

Part One

"Howard Cut His Teeth on a Camp Spoon"

The first time I saw Howard he was working in a heap of shavings the color of his blond hair, trying to make my bedroom door at his mother's house close more easily. He wore the colorful clothes of a hard-working logger, red-checked woolen shirt, Malone breeches, laced boots, and a tall Scotch cap pushed far back on his head. He looked much like a high-school boy dressed for a part in a class play. (I think the lumberjacks still call him "Old Man Anderson" because he looks so young.) I remember well his sudden boyish smile when he looked up from those sharp-smelling shavings that first day. His deep-set gray eyes took my girlish fancy in a sudden surge by their brightness and their sparkle.

"Hello," he said. "Mom told me your name's Julie Benson. May I call you Julie? You'll be sort of like one of the family living here. Mom mothers all the teachers."

"Sure. You may call me Julie. I'll call you Howard, then. You *are* Howard, aren't you?"

Because there are no trains except those carrying ore and logs out of Wakefield, Howard, after a few weeks' fast-developing comradeship, drove me to my home one weekend. When I introduced him to my folks as a logger, they were quite astonished. I am sure they expected all loggers to be Paul Bunyans, if not in accomplishments, at least in bulk. On the contrary, Howard is Danish; tall and very slender—like his father—not heavy-set or powerfully built. His voice, which can be heard all over a section, completely compensates for what he lacks in immensities of physical stature. Howard really does two men's work every day, has an enormous appetite for unusual foods, and smokes cigarettes incessantly.

"At Lawrence I tried smoking a pipe, like my dad," Howard told me. "But I always went back to smoking cigarettes. When I came home to log for Dad, I found that filling a pipe was too much trouble when it was ten degrees below in the woods. My lumberjacks all chew snoose or smoke cigarettes in the winter. In the bunkhouse some light up their pipes. Dad used to say he'd never hire a jack who smoked cigarettes. He'd turn him down just on the supposition that cigarette-smoking jacks could be nothing but weaklings. He broke down and hired me, though."

"Well, so you're a Michigan lumberman?" my own dad asked Howard during his first visit with my logger.

"Not exactly a lumberman," explained Howard. "You see, in our section of the country we call a man a lumberman when he sends logs through his own mill and then sells the lumber product. A logger, like I am, buys timber, operates logging camps for lumberjacks who cut the trees, and then sells and hauls the logs to either lumber or veneer mills. Loggers cut other forest products, too, like mining timber, lagging, and ties. These products are delivered to the mines and railroad landings. To boil it down, a lumberman deals with lumber, while us loggers handle *many forest products.*"

"Logger or lumberman, you're in a clean and interesting business, I'd say." And logger or lumberman, my father liked Howard from the first time they met; not, I'd say, because of Howard's work, but because Howard is a man's man.

"What does your father do?" my dad asked.

"My father was a logger, too. He died as a result of a woods accident. Since then I've been finishing up his logging job and taking care of my own camp, too. Keeps me kinda busy."

My mother, who had survived my high-school infatuations

and college romances, sighed when we were alone. "An architect, an interne, a history teacher, a chemist, and now a *logger!* Who next? Your Grandfather Clark logged for a good many years before he settled down to farming. Watch your step, young lady! I've seen what the life of a logger's wife is like. I remember real well the song my father sang to tease me when my beaus came to call. It goes like this:

Oh, loggermen are a crooked breed,
A-judgin' by their deeds,
Of stealin' one another's white pine logs
'N makin' up big lies.
Cussin' lots 'n boozin' up is how they exercise.
So if you want to settle down
In comfort and in joy,
My best advice to you is this:
"Marry an honest farmer boy."

"But, Mother," I protested, "you don't really mean that. Logging has changed since Grandfather's day, and you must admit that loggers are nice people, especially Howard. He's so conscientious—so nice! His smile! Have you happened to notice that gleam of reddish gold that makes his blond hair so pretty? I think Howard's rather handsome. In fact, I like Howard a great deal."

And before Howard's first visit at Shell Lake ended, Mother liked him as much as Dad and I did.

It took me several weeks to adjust to teaching in this remote Upper Peninsula community. To be able to pronounce the names of the students, I absorbed quick courses in Finnish, Polish, and Italian from a teacher who had taught here several years. A great deal of my time not consumed by teaching, teachers' meetings, or teachers' teas I spent at the Inn, appeasing my violently increased appetite with Cornish pasties or Italian ravioli—delicious concoctions, new to me, which I have since learned to make.

Evenings when Howard stayed at camp, I visited with

his mother, a tall, reddish-haired woman with friendly brown eyes. When I asked how Howard happened to be a logger, she said, "Oh, it was just natural for Howard to like logging. Why, he cut his teeth on a camp spoon. All during his boyhood he went to camp with his father. The lure of tall timber which kept my husband Louis engrossed in logging all his life also fascinated Howard. In his father's camps he learned to enjoy the gamble and excitement of logging as an occupation, and being with his father, he acquired through experience a knowledge of logging procedures for which there is no easy ten-lesson course. Louis learned logging that way from Mr. Jenson, a lumberman who came to the United States from Denmark when white pine still covered most of Michigan. When I met Louis he was working for Mr. Jenson as walking boss of his lumber camps around Gaylord, Michigan."

"Wasn't it strange that you should meet a logger?" I asked, wondering how a busy man who spent weeks and months at a time in logging camps found time to court.

"My meeting Louis was not so strange as you think. I worked in my father's law office, where I compiled his papers and did his secretarial work. When Mr. Jenson established a lumbering business at Gaylord, my father took care of all his legal matters. One day Mr. Jenson and Louis came to the office together, and that's how I met my husband."

As we thumbed through a collection of family pictures, clippings, and photographs of logging operations, Mrs. Anderson told me this casual acquaintance became a friendship which culminated in their marriage in 1909. When Mr. Jenson, a foresighted man interested in new business ventures, followed the speculators to northern Michigan and established lumber mills at Ewen and Tula, Louis Anderson came with him to boss the logging operations in camps along the Ontonagon River.

"In nineteen-thirteen," Mrs. Anderson related, "while Louis was still working for Mr. Jenson at Ewen, Howard was born. By the time Howard was of school age, his father had bought timber of his own and was logging around Choate and Paulding, sending his logs down the Ontonagon River to Jenson's mill."

Howard's mother fingered the pictures of big log loads and told me how, shortly after her husband came to Ewen, he acquired the name of "Hungry Louie," not because of his continuous appetite or his lean, hungry look, but because of his insatiable demand for logs, logs, and more logs. Isabel had heard that he would appear at a sawyer's strip puffing on a corncob pipe—"Louis always smoked some kind of a pipe to which he continuously held match after match in an effort to keep it lighted," she said. "He would count the sawyer's logs and then demand, 'More production, men. Gotta have more production. I see you didn't cut that last log outta that good pine. Can't let sound timber lay. Can't make a livin' sittin' on a stump. Gotta have more production, men,' he'd say.

"No matter how hard the jacks worked for him, Louis always made the same challenging comment. Jacks began to call him 'Hungry Louie' because he wanted faster cutting, more logs, and increased profits. His friends called him Hungry Louie with a real esteem for his tireless logging activities. Logging kept him continually gaunt, even though his appetite was enormous. The name 'Hungry Louie' thus became even more appropriate."

Like other loggers, Louie Anderson bought timber that was the most profitable logging chance, even if such policies meant moving frequently from place to place. Moving often was disagreeable to Howard's mother; but she accepted the circumstances, for she understood that it was necessary for a logger to go to the timber. The obvious reason led the Andersons on and on. When their timber at Choate was logged, they moved to Moran on the Charles River, where Louie operated three camps and in a few years accumulated a small fortune.

Mrs. Anderson recalled that the only way she could keep

her lively and energetic son amused during the time he was not in school or with his father in logging camps was to read to him or give him embroidering to do. When I would not believe that Howard, a logger, could possibly have done embroidering, she produced a number of doilies and dresser scarves adorned with French knots and lazy-daisy stitches. All this girlish activity ended abruptly when Howard was given a pony. The problem involved that of the linen taking on the smell of the pony and the barn. I have often wished Howard had learned plain sewing instead of fancywork; then he might have saved me the trouble all these years of mending the three-cornered tears in his breeches, or putting new pockets in his pants which he uses as tool boxes. Alack! It was my fortune to marry a walking tool box who would not mend—even his ways.

While his father acquired a fortune, Howard began his formal logger's training in his father's cookshanty. When he was thirteen—small for his age but with the ambitions and energy of a much older boy—he became official cookee for the summer and was listed on Louie's payroll as receiving one dollar a day and his chuck. Howard slept in the upper bunk in the office. From daylight until dark, he swept the plank floors, dried the pots and pans, cleaned the lamps and filled them, and split wood to fill the wood rack by the cookshanty stove. If he could stay awake after supper, he listened to the hell's-a-popping stories of the lumberjacks. His mother declared to me, as she told the story, "I'm sure the wild tales he heard were mostly the lumberjacks' versions of the Western stories from the bunkhouse supply. But they were surely thrillers to Howard."

With the idea of retiring from the lumbering business in mind, the Andersons moved to Bessemer at the western end of the Upper Peninsula. But log-hungry Louie could not remain idle and see tall timber blown down, even if he had decided to quit logging. Accordingly, he bought a remote section of timber, a locomotive, and Russell cars. He built a railroad to the timber, constructed a set of camp buildings and a landing, and entrusted the logging to a foreman. Howard describes these events as "the year when Dad lost his shirt and tightened his belt just in time to keep from losin' his pants, too." The foreman dissipated Louie's profits, left bills unpaid, and finally skipped the country, forcing Louie to take to the timber in order to pay his bills and replenish his fast-dwindling cash supply.

Howard, characteristically, was always more interested in logging than he was in following some profession. His mother told how he humored his father's wishes and entered Lawrence College in Appleton, Wisconsin, to study law. "Howard enjoyed debating and found his studies not at all difficult," she told me while we sipped hot coffee in her kitchen. "But during the years of depression following 1930, he withdrew from Lawrence and went to Detroit to live with an aunt in order to continue his education at less expense. Finally, Howard came home to Wakefield to log with Louis. The struggle to make a living by logging was great, but together Howard and his father did it.

"One summer Howard hauled ties for his father and unintentionally dumped them in the wrong place. Louis rode him hard about it; to make things right, Howard proceeded to move them. He slipped and fell while carrying a tie. It smashed his face, flattened his nose, and fractured his skull. He was partly conscious when two railroad men found him and took him to the hospital, where he spent days of agony and nights of loneliness while his nose and skull mended. The price we received for those ties was small in comparison with the hospital bill and the personal discomfort Howard suffered for trying to satisfy his father."

The two Andersons had more than their share of tough breaks and financial lickings, but they had an irrespressible spirit which could not be conquered. They fought their way through years of discouragement with logger tenacity. It seemed that as a team they were unbeatable. But the end came swiftly.

The seasons come and go, in an eternal change, but always with their gift of beauty. So it was in Louie Anderson's logging camp. Winter came with its spectacle of enormous and fantastic snow shapes glittering in the sunlight. Late in January of 1938—seven months before I came to Wakefield—came the worst storm this region has ever experienced. Snow fell in heavy flakes for two days and nights without stopping. The wind blew, the snow drifted, roads became impassable, county snowplows were stalled behind tremendous drifts, schools were closed, business places were inaccessible, everyone was confined to the snugness of his own home, and the men in logging camps had three days in the hay, except for those trying to keep the woods roads open.

Mrs. Anderson had not seen her husband for several days and began to be worried when the storm became bad. From the foreman she learned, on making inquiry, that Louie had worked continuously during the storm, trying to keep his woods roads open, but apparently had never loaded another log after the storm. Howard was busy digging his way out along his own logging roads and learned the fatal truth when he came home. Tired from so many extra hours of snowplowing, Louie Anderson lost his footing on the icy road, slipped, and was crushed between a truck wheel and the snowbank. He was breathing with difficulty when his men gently laid him on warm blankets arranged on a sleigh. They covered him carefully. Behind a fast team they sped over the logging roads to the highway, where they transferred him to a car. They took him directly to the Wakefield hospital.

For a number of days Mrs. Anderson hoped for her husband's recovery, doing all she could to comfort and solace him. Howard, with a son's true devotion, stayed with his father until the end. Louie Anderson, strengthened by his family's loving concern and the best of care, struggled to live; had the drugs of today been available at that time to help him fight the pneumonia which developed, he would, perhaps, be alive now. Instead he died—a logger in his prime, active, well loved; he was a man I wish I could have known. Many remember Louie Anderson as a fun-loving person, generous, sincere, and honest. He was a deeply respected logger of indomitable courage in the face of any odds; and he had given a priceless heritage to his son Howard—a way of living above the reproach of any man.

"Scat, You Coyotes"

Howard was forced to do two men's work in the months following his father's death. There was no choice except to finish the logging Louie Anderson had started or else lose the investment in stumpage, buildings, roads, and wages he had paid the lumberjacks. Howard tackled and finished his father's job profitably, but not without a struggle. My admiration for him, during those months I watched him fight against odds, grew in proportion to his resistance to defeat.

As busy as his logging jobs kept him, Howard finally took time to ask me for an honest-to-goodness date—which I had been wishing he would do. From that time on we did together the usual things teachers are allowed to do and still be considered fit people to teach the children of parents who (as I sometimes peevishly speculate) do all the things it is not considered proper for teachers to do. Going to the theater and school dances and having a sandwich together was the extent of our associations until Howard suggested, "If you enjoy hiking, you might like to go with me to my dad's camp, Julie." During the week Howard stayed at his own camp, but on Saturdays or Sundays after his own crew was paid, he had to find out what had been done at his father's camp and to plan the next week's work for that crew.

The first Saturday I decided to go to camp with Howard turned out to be a perfect, sunny fall day, the kind that makes one feel vivacious and full of aspirations. The leaves were beginning to turn color, and the air was refreshingly cool. There had been several days of heavy rainfall, but I thought nothing of that when I polished my boots and put on my best riding breeches. When I was ready to go, Howard smiled sympathetically and asked, "Don't you have some *old* boots and pants?"

Indignantly I asked, "What's wrong with *these* pants?" I was hurt to think my appearance did not suit him. What I did not know then was that he heartily approved of my appearance. He was concerned mostly about my comfort. If I appeared in those same breeches and boots today, he would say, "For Pete's sake, get into some different riggin'! Do you think you're goin' to ride the skiddin' team? There isn't a bridle path closer than Mercer, and I'll be *hanged* if I'll drive that far just so you can strut in that getup." (Most loggers really use *very strong* language. Perhaps unprintable. It is one of the things I have learned to accept.) Strange as it may seem, that outburst would indicate the same sincere concern he expressed so mildly *before* we were married.

Howard was dressed in warm, comfortable clothes. He wore rubber boots made for tramping through ankle-deep mud, loose-fitting pants with the cuffs rolled up, and a bright-red plaid shirt. His pockets bulged with a supply of cigarettes.

I found out the hard way that fashionable, ankle-high riding boots and trim, tight-fitting jodhpur pants are not appropriate to wear when tramping over forest trails and hiking down old logging roads. We drove to Jack's Spur, parked the car at the landing along the railroad, and started into the timber. The logging road, at the beginning, was hard and without deep ruts. I kept up with Howard, swinging along beside him, taking as long strides as my tight breeches would allow. This was the first time I had ever walked in virgin timber. The enormous hardwood trees, their extreme height, and the denseness of the timber were a thrilling sight to me. I tried to imagine what white pine forests must have been like, years ago, if these remaining hardwood trees had been scorned by early lumbermen. Those virgin stands of white pine, thousands of acres, have been described as "a howling wilderness." Those magnificient forests—the haunts of Hiawatha—penetrated only by cruisers, early pioneers, and hardy lumberjacks, who left blazed trails, wagon ruts, and skidding roads to tell of their passing, were, in truth, a howling wilderness. The virgin hardwood we passed through stood just as it had been in those days; and progress—towns, railroads, and highways—by presently approaching without real encroachment, brought mankind close to a sanctuary of beauty, strength, and magnitude.

Bright red and yellow leaves covered the road with a gaily patterned carpet which disguised the treacherous six inches of mud on the last two miles to camp. I could not keep up with Howard, and he stopped now and then for me to catch up to him. I slipped into ruts ankle-deep with sucking mud, and my freshly polished boots became caked with it. Even my breeches were covered with the chocolate-colored goo. When I tried to go around the worst muddy places, sharp twigs and low cedar branches caught at my hair and scratched my face. I became short of breath, my legs ached, and I turned my ankles innumerable times. Even a seam in my breeches burst when it could no longer stand the strain of all the acrobatics I performed trying to keep right end up on the slippery road and not too far behind Howard.

Once, while we rested, Howard's strong arms embraced me. The autumn sun suddenly became warmer, and the sky seemed bluer. The grosbeak's call sounded sweeter, and a squirrel's beady eyes became more friendly. I knew Howard spoke truthfully when he said, "I love you more than all the world. I want you to marry me." He was looking down at a girl whose hair was full of bits of colored leaves and blown into tangles and snarls, whose blouse was torn, and whose breeches and boots were caked with mud. His proposal of marriage was certainly not aroused by glamorous circumstances.

"We both seem to like the same things, honest and sincere friends, being in the woods, and—let's hit the trail together. What do you say?" Howard urged.

"For weeks I've wondered how you felt about me, Howard. I love you, too," I confessed, without being coaxed.

I never expected a proposal of marriage while nearly mired in a muddy logging road. Neither did I expect to accept such a proposal with so little deliberation. But I did, and being a logger's sweetheart and wife proved to be exciting, thrilling, and difficult, at times, but always worth any sacrifices.

We were reluctant to leave that enchanted forest spot, but because Howard had much to do we continued our hike through mud and muck on to camp. I puffed and struggled along behind Howard. Outwardly he showed little sympathy for me. Several times we rested for as long as five minutes, but Howard, who never seemed to become fatigued, would start down the road again before I had fully recovered from the past half-mile. I dragged along, frequently wishing I had never come.

In a clearing along the road I saw a tarpapered building. I thought we were at last coming to camp, but Howard said, "It's just a shacker's place." The shacker must have seen us coming for he appeared in the doorway and greeted us in a most unusual manner. His gray hair was tangled and matted, and his whiskers were the regulation Santa Claus length. Clad only in a suit of dirty long-sleeved and long-legged underwear and waving a bottle of whisky in one hand, he began to stagger down the muddy road barefooted. Suddenly he must have discovered that, for all my mannish attire, I was a woman, for he stumbled back into the shack. Howard apologized for the shacker's wild antics by mumbling something about how he guessed the shacker was still celebrating the Fourth of July. We continued on to camp. The sight of the low tarpapered camp buildings—long patches of black velvet stitched with frayed heather-gray smoke threads to a multicolored autumn quilt of hardwoods, bottle-green pine and hemlock, and gentian-blue sky—was certainly welcome to me.

The sweaty and visibly fatigued jacks, coming in from the woods for their noon meal, eyed me suspiciously; I supposed because of my crimson vest and torn and muddy breeches, but Howard explained that it was just because girls did not often visit logging camps. I was too tired to care what the jacks thought or what I looked like and dropped into a bunk in the office to wait until the jacks had finished their meal.

When the last man left the cookshanty, the cook set places for Howard and me. With heaping dishes of many varieties of food before me, I hardly knew what to eat. I stuffed myself and then somehow managed to eat a slice of apple pie and a huge piece of chocolate cake. I waddled back to the office to wait for Howard to "turn out" the men. When he came in to get his scale rule and cigarettes, instead of going with him to watch the sawyers as I had intended to do, I curled up in the office bunk where I slept until late afternoon. Howard came back to camp for coffee when the sun was dropping down behind the treetops. With a long list of groceries the cook needed tucked in his pocket, and a mental list of many things to be done before his next trip, Howard suggested, "Let's leave before it gets late and cold. I think there'll be frost on the ground in the morning." We started back along the logging road to the car.

The sun disappeared quickly, and the sky became a rosy gray. As the purple shadows of the trees closed in upon us, I kept near to Howard. By disregarding the mud, I made better progress than I had coming out. Finally it became so dark that Howard hunted around in his packsack for his flashlight. As the light flooded the road ahead, we caught sight of a pair of coyotes sneaking into the timber. I shivered, my teeth began to chatter, and beads of nervous perspiration made my hands clammy. Howard laughed at me for being frightened, but he walked beside me and put a protecting arm around me.

Even more comforting was his unusually low and affectionate voice, telling me, "Those coyotes are more afraid of us than you should be of them. A pack of them might attack a person if they were starved, but not if a fellow has any kind of a light—a lantern, a flashlight, a burning pine knot, or a campfire. They are afraid of fire. But to be real safe," he tapped his packsack, "I have a revolver too. Never had to use it yet except to shoot a horse that fell and broke a leg." We both laughed at this quip. Laughing relieved my nervousness.

When the green eyes appeared again in the range of our light, I said, "Scat, you coyotes," and felt much better. The coyotes escorted us all the way to the highway. As fast as we walked, they slunk along in the concealing darkness of the timber. Occasionally their eyes gleamed in the radius of our light. When we reached the landing, I climbed into the car with great relief and slammed the door hard.

At home I peeled off my muddy, torn clothes, left them in a heap, and crawled into bed. All night long I dreamed about bowls of baked beans, tall trees shedding crimson leaves, grasping branches that would not let me move, coyotes' fiercely gleaming eyes, and the strong encircling arms of a blond man wearing a lumberjack shirt. My respect for Howard had become a declared love.

After I recovered from that first hike with Howard through the mud, and stopped dreaming about coyotes' eyes gleaming in the dark, the memory of discomforts became minor. I remembered only the extreme beauty of the forest in fall colors and what an exhilarating experience it had been. I bought a pair of soft leather hiking boots, several bright plaid shirts, and a pair of warm brown slacks, a size too large but comfortable. (On the same shopping trip I brought home a Gage hat—a fashionable creation, but not to be worn on hikes through the timber or down logging roads—which saw the light of day only once before I gave it away.)

Nearly every Saturday morning while my friends slept, I went with Howard to his camp, where I hiked along the woods roads, bright with colored birch and maple leaves.

While Howard was busy scaling loads and counting the logs in the sawyers' strips, I walked alone except for Tex, Howard's friendly dog, who was always at my heels when he was not chasing squirrels or pouncing on the quivering branches of newly felled trees.

Tex, who looked exactly like one of Lassie's pups, was the dog Howard had bought for a dollar from a Texaco gas salesman—hence, the name Tex. He had carried the puppy home in his arms. Then, Tex was just a fluffy puppy that would appeal to any fellow who liked dogs, horses, the out-of-doors, and logging. Now, Tex was a full-grown dog, with shiny fur, a truly beautiful collie. He, like Howard, loved to be in the woods and enjoyed the excitement of logging. All day long he followed the sawyers from one tree to another; and as each tree was felled, he would bound to the top branches while they were still quivering from their sudden descent, grab them in his teeth, shake them as any other puppy would shake an old slipper, and growl. As soon as the branches settled, he would go on to the next tree being felled, sit patiently while the sawyers made their notch and pulled their saw back and forth to fell the tree.

When the crack of the falling tree sounded, when "Timber-r-r" warned all others of a falling tree, and the sawyers stepped to safety, Tex would bound quickly around the stump down the length of the falling tree to the top branches and ferociously shake and growl at them. Playing in this manner all day, he would return to camp so tired that he was content to lie anywhere and just pant.

Tex, although he spent a great many hours following me or the sawyers, was a one-man dog. Even I could not abuse Howard in a friendly manner; for when I did, Tex would spring between us with just enough growling and nipping to let me know that he meant to interfere. I am sure if further defense for Howard's sake had ever become necessary, Tex would have put up a good fight in his behalf.

Of course, Tex was my friend, too. He had occasion to

prove it, much to my satisfaction and some surprise one afternoon. A horse Howard had been doctoring was getting well, but he needed exercise. Because the day was warm and sunny, Howard suggested that I untie Cap and lead him out into the sunshine. I never frequented the barn because all the horses were lively, the barn small, and I was not interested in being a target for a horse's hind feet. More than one teamster climbed the oats box just in time to avoid them. As I was not adept at climbing an oats box, I just stayed out of the barn—except on the day when I tried to give Cap an outing. He was in the first stall in the barn on the left-hand side of the back door. All I had to do was slide in between the horse and the barn wall, reach up to the halter, and untie the thick rope fastened to the oats box.

Tex was behind me, and everything went well until Cap began to crowd me. Logging horses are big, heavy, powerful horses, bought because they can pull hard and endure strenuous hauling and skidding of logs. When Cap began crowding me against the barn wall, my heart pounded so hard I was sure it was right in my ears. I tried to push him away; but he crowded me more, and I found that I could not breathe. I began to get dizzy and thought that if Cap did not break my ribs, the pounding of my heart would. Tex then proved that he had a dog's love for me and some horse sense as well. He apparently knew that I was in a dangerous position and frightened, for he leaped at the horse's heels from my side. With several convincing nips, he made Cap swing his rear hard up against the inside stall wall, giving me space to get out of my crowded position—which I did in no time at all.

Breathing rapidly, I staggered out into the sunshine. A feeling of immense relief engulfed me, and the pounding of my heart subsided. I could show my gratitude to Tex only by patting his head. It was all he expected, but certainly little in return for his being quick to sense my dangerous position and act as he did. After that incident he reversed his tactics when we were in the woods and nipped *my* heels when he thought I was in the path of the skidding team.

Those walks through the virgin timber on Saturdays with Tex were the first of many such hikes, but I have never ceased to be thrilled and awed by the extreme beauty of streaks of sunlight shining through tall trees, in silence so deep that the falling maple leaves can be heard rustling as they float down and touch the earth. To me, looking up, the trees make an archway to the cathedral in the realm of God. No man-made thing this archway, nor the music that sounds from the cathedral! Just the wind blowing through the tiptop branches of the trees—deep and throaty music sometimes, and sometimes high and piercing, yet always beautiful and always there for those to hear who care to listen.

Only my hunger lured me from the tall timber back to camp. After eating a big meal, I would wait in the snug little office while Howard figured the lumberjacks' accounts and wrote their checks, which he told me would be spent mostly for whisky. Every Saturday more than half of the jacks went down the road to Hurley, to return on Sunday or early Monday morning, disgustingly broke, pathetically sick, and eager to earn a few more dollars to be spent in exactly the same way the next time they drew their pay.

Waiting in the office while Howard figured scale sheets and strip counts was seldom tiresome for me. I fed fat chunks of wood to the greedy barrel stove and read magazines from the bunkhouse. When the last scale sheet was added, Howard would climb down from his high stool behind the counter to take a handful of checks to the jacks waiting in the bunkhouse. The late afternoon sun cast a long thin shadow of his slender, slightly stooped figure in front of him as he walked slowly down the road.

When he returned to the office, the sun shone on his red-whiskered face and exposed lines of worry and fatigue that I had never noticed before. But he would smile broadly as he swung open the office door. "Well, Julie, let's call it a day," he would say. "Are you ready to go back to Wakefield?" He would check the fire and clear the counter of his papers, and we would step out into the fresh cold air.

As we drove into Wakefield the sun usually flamed low in the west, sinking fast from sight, and the warm, bright colors of the trees' foliage would become lost in the evening darkness. The rugged hills, jagged ore piles, and towering mine shafts became simple silhouettes against the darkening blue of the sky. Tired, but satisfied at having completed another week of good logging, Howard would look at the last of the brilliant sunset and quote an old Danish saying. "'Red sun in the morning; sailors take warning! Red sun at night; sailors' delight.' Tomorrow ought to be a nice day, Julie."

And I would say, "Today's been a nice day."

We would arrive in Wakefield as the last faint rays of sunset disappeared and yellow shafts of light from the homes broke through the early evening dusk. Red tail-lights on logging trailers glowed softly in the early darkness as drivers headed for home. Lights on distant mine shafts flicked on, and stars would appear bright in the heavens above this rugged mining and logging region. Howard's arm rested lightly around my shoulders. Expertly he would round the corner to his mother's home; and, as I leaned his way, he would kiss me quickly on my cheek.

Drifting Snow

Snowfalls in this part of Michigan are measured by the foot. The sparkling star-shaped fluff is always clean, because no sooner is one snowfall covered lightly with soot than more gently falling snow disguises the sooty grime. People do not bother to shovel the snow from their walks, because eventually there would be no place to put it. Pedestrians share the streets with other traffic rather than continually shovel snow or take chances on getting lost during zero weather in a maze of shoveled walks confined by walls of snow too high for one to see over. Logging roads and the highways become increasingly narrower, and the banks on either side become higher with each snowfall and successive snowplowing.

As well prepared as Upper Peninsula cities and counties are for removing snow from streets and highways, usually once during the long, cold winter so much snow falls and the wind blows it into such huge drifts that several communities are snow-bound.

The first time I ever was snow-bound in Michigan was most pleasant. An announcement came early that morning over the radio telling us that no schools on the range would be open until the plows had cleared the roads. Without bothering to look outside, I pulled the covers closer around my shoulders and slept for several hours more. Hunger aroused me. It was still storming. The trees and the roofs of neighboring houses were already so heavily covered with snow that it seemed to me the weight of another snowflake would break the trees or cause the roofs to collapse.

In the blinding storm I crawled through one drift to another, sensing rather than seeing my way to the Inn, where other teachers were already shaking the snow from their clothes and exclaiming, "Did you ever see anything like it? It took me three-quarters of an hour to crawl here when I usually walk this distance in ten minutes." Someone else said, half-suffocating, "Somebody please get this snow out of my ears and collar before I turn into a snow man!"

The high drifts forming around the Inn gave us the feeling of being completely isolated. The music from the juke box kept us merry. The kitchen was well stocked with any number of good things to eat. Someone started a bridge game, while others of us danced to the "Beer Barrel Polka" and other current gay and noisy tunes. For us teachers the day passed quickly and pleasantly. But for Howard, who had worked at camp all night plowing snow from the woods roads, it was a nightmare. I returned to my room late in the evening when the storm had subsided. Sometime during the night Howard came home to his mother's, where I roomed, for a few hours of undisturbed sleep.

The next day, a Saturday, when the storm was over, I drove to camp with Howard. The main highways were plowed, but most of the side roads remained smooth, satin ribbons of snow, broken only by tracery of deer and rabbit tracks. The road from the highway to Howard's camp had been plowed, and the men were already clearing skidways and digging out the log decks. The low, black camp buildings lay deep in snow. Only the highest of the bunkhouse windows and the one by the counter in the office could be seen above the drifts. Enormous inverted bowls of snow buried stumps and brush piles. Tremendous drift formations and branches bent with ponderous loads of snow gave an uncanny quality to the fantastic shapes everywhere around.

That day, at Howard's suggestion, I tried to use a pair

of snowshoes to explore the beauty of the mountainous drifts. That was the last time I ever had the courage to put on those treacherous contraptions. The only possible means of making progress through the deep drifts here is to wear snowshoes. To please all who have special requirements for winter travel through the timber or over snow-covered trails, there are several kinds of snowshoes. Some are long and narrow; others are short and wide. Some are sturdily made and heavy, while others are very lightweight. My requirements were merely that they keep me on top of the drifts. At that they failed miserably. I thought the snowshoes were useless to me because they were the wrong type; and I said so. Howard reserved his opinion for later. He helped me fasten the laces over my boots and said, with a wry smile, "Don't go too far. You'll get tired coming back."

Walking with frameworks of wood and woven thongs laced to one's feet is awkward even for experts. Any snowshoer has to be slow and deliberate in placing each shoe apart from and ahead of the other shoe until the action becomes natural. Before I had taken four steps, I was sprawling in the snow. I had tripped myself. At that moment the cook appeared in the doorway to watch me. Between bursts of laughter he offered suggestions which sounded as if they would be helpful if I had been in a position to try them.

Howard watched from the office window while I struggled to get up. Finally, I regained a standing position and my composure. Confident that the snowshoes were governable, I started out again. (Howard uses them with ease. Every winter he travels many miles through the timber over the drifted snow faster than he could travel the same trail in summer, when it is necessary to crawl over logs and go around brush piles and swampy places.) But the "friends of the woodsman," with which Howard travels so well, threw me again into the deep snow.

With my mouth full of snow, my face stinging and wet, and my legs twisted into a painful position from which I could not untangle myself, I signaled wretchedly for help from Howard. The mischievous twinkle in Howard's eye as he helped me to my feet and unlaced the ankle straps showed me how much fun he was really getting out of my troubles in trying to tame a pair of snowshoes. They are, in my opinion, a very smart device for any practical joker. Howard just remarked dryly, "Seems to me you are the wrong type for snowshoes. Ya know, for you skiing's like hatching eggs—requires a lot of sittin'."

"Well, I'd much rather hatch an ostrich egg than walk another step shackled to snowshoes." And for all our good-natured kidding, I've never had on snowshoes since.

Howard helped me brush off snow, hung the shoes on the office wall, and returned to checking the sawyers' log count. After a day and a half "in the hay," not being able to saw logs because of the storm, most of the sawyers were restless and thirsty. The number of jacks going to town was greater than on other Saturdays. While Howard figured his payroll and Apie (Howard's cousin who drove a tractor for him) and I soaked in the warmth from the stove, the sky darkened and more heavy flakes of snow began to fall.

As Howard wrote the first check, he glanced at the calendar which hung at the end of the van counter. It was January twenty-ninth—the date which had been encircled by a green marking crayon. He looked for a long time out at the swirling, drifting snow; then he swallowed hard, and silently turned back to writing checks and figuring.

Apie's eyes had followed Howard's to the calendar and then turned back to stare at the floor between his knees. He said quietly to me, "Howard's dad died a year ago today. It was storming just like it is now on the day Uncle Louie was injured." The only sounds were the scratching of Howard's pen and the crackling of the fire. The small office was filled with the grief of two men. The fire burned out; the last check was torn from the checkbook, and Howard took them to the jacks.

When we left camp after supper, the wind ceased and the snow stopped falling. Branches broke with the weight of the snow. The cracking noises sounded loud in the still, cold air. Whiffs of wood smoke drifted down to scent the air. The sky, which had been gray and menacing, became bright with stars. Moonlight masked reality; it transformed the camp buildings and the surrounding drifts and snow-burdened trees into the habitat of Paul Bunyan ghosts.

All winter Howard divided his days between two camps, and I spent weekdays teaching. It was while he was icing the woods roads for sleigh hauling at his dad's camp that he caught the lace on his mitt in the belt of a motor and broke his right arm in two places.

Icing roads was done at night in preparation for the next day's hauling. The accident happened at one o'clock in the morning. With his broken arm cradled inside his shirt, Howard walked three miles to Jack's Spur, where he had left his car, drove to his own camp to get his foreman and crew started for the day, stopped at the hospital to have his arm X-rayed and put in a cast, and arrived at his mother's home in time for breakfast with me before I left for school. He had traveled a distance of fifty-seven miles. His ashen face showed the pain he silently suffered. That he had driven so far with the use of only one arm seemed unbelievable. Truthfully, I would have doubted it if I had not had previous proof that he could drive a car with one arm.

One Saturday toward the end of the winter logging season, when I was free of teaching duties, I came to camp with Howard to hike along the woods roads. We had just arrived and been in the office only a few minutes when the cook pushed the door open and blurted out, "I'll take 'fire time,' boss."

"Goin' down the road today, Joe?" Howard asked.

"Yep, gotta buy new shirt and pants."

"Don't you need a tooth pulled, too?" joked Howard.

"Gotta buy shirt and pants," repeated the cook sullenly as Howard wrote his check. He made no further explanation. Howard handed him his check. The cook folded the check carefully, put it in his shirt pocket, and slung his packsack over his shoulder. The door banged shut behind the most important man in the crew. And no one knew his real reason for quitting.

There were shirts and pants of all sizes and colors under the counter. "Why didn't he buy a pair of pants and a shirt here instead of going to town for them?" I asked.

"Because when jacks quit, the excuses they give are neither logical nor real. If they ever return to the same camp, it is without that bad tooth pulled or without the new pants and shirts.

"Well, your cook is gone, and there are thirty men to feed. What are you going to do?" I asked with some pretty sad concern. I thought Howard had already had enough bad luck without having his cook quit, too.

The next man who came to the office was Baldy. He was a good-natured fellow of all-around accomplishments who calmly listened to the news about the cook quitting. Baldy was ready to do what he could for Howard in this bad fix.

"I'll throw some spuds in a kettle," he blandly stated, "open a gallon or two of peas and peaches, roast a chunk of beef, and stir up a little gravy." *That* was all there was to preparing Sunday dinner for thirty hungry jacks. "But," he maintained, "the jacks will *all* quit if they don't have pie. No pie, no crew, so—no cash!"

Trying to be helpful, I rashly said, "I can make pies."

That afternoon was the first and last time I ever ventured behind the pie board in the cookshanty. Pies? I could make good pies—one at a time. "For how many men did you say, Baldy?"

"Thirty," he said. I thought there was some humor in that one word, but I had committed myself to making pies, and I silently left for the cookshanty. In a matter of seconds I had agreed to make at least five pies. "Should they be made one at a time, or, all at once?" I asked myself.

Surveying the shelves concavely curved with the weight of every imaginable kind of seasoning and supplies, I tied a big flour-sack towel over my snow pants and looked for the measuring cups, spoons, and a suitable bowl. The flour, lard, and a bowl were not hard to find, but there were no measuring spoons or cups. Conveniently, there was enough (I thought) stewed apricot sauce for five pies. A trip to the office to ask where the measuring things were revealed only the information that camp cooks measured flour in their fists or any kind of cup, and that the lard was scooped in "by-guess-and-by-gosh." Puzzled and wondering what these pies were going to be like, I returned to the pie board to experiment with the by-guess-and-by-gosh method.

I dumped flour into the bowl. There seemed to be too much. I put some back into the bin. Then the flour in the bowl appeared to be not quite enough for five pies, so I scooped more flour back into it. I scooped out and put back, until I was dizzy. Who invented the by-guess-and-by-gosh method? I had my suspicions. Finally I realized I had to make up my mind about the amount of flour needed for five pies, or else never produce them. No pies, no crew, no logging, so—no cash! From dozens of long-handled spoons and ladles hanging on the pan rack over the stove, I selected a spoon and scooped lard into the bowl. Adding salt and rolling the flour and lard between my hands, as my mother had taught me to do for a flaky crust, was the easiest part of making five pies.

With ten lumps of dough piled before me, I floured the board and reached for the rolling pin. Its weight kept it anchored solidly to the board. When it did start to roll, it dropped on my foot with such force that I thought my toes were broken. I pictured that rolling pin as the output of some blacksmith shop, where it had been loaded with lead. I managed to get it back on the board and rolled the first pile of dough.

The queer shape that developed under the heavy rolling pin completely amazed me. It was as lopsided as I myself felt now, standing on one leg to ease the pain in my nearly crushed foot. I patched and pushed the queer shape into a pie tin and hurried to roll another crust. All five crusts were as far as possible from being round. I would have cut off the scalloped edges, but there was not enough dough for that. I filled a pie with the sauce. *This* was my day for frustrations —what I thought would be plenty of sauce for five pies just nicely covered the crusts after I ladled sauce from one crust to another to make them all equally full.

Some bird from the office twittered the tale to the jacks in the bunkhouse that a schoolteacher was trying her hand at baking pies for the camp crew. Curious jacks came continuously, by twos and threes, to the cookshanty for coffee during the entire afternoon. Each time the door slammed shut on the departing jacks, there was a short interval of silence while they walked back to the bunkhouse; and I sighed in relief to have the smiling men gone. Then a low rumble of laughing and talking coming from the bunkhouse would spur me on to completing my job before another delegation of jacks came on an investigation tour.

The visits of the lumberjacks made me nervous. It was hot by the stove I had stuffed full of the biggest sticks of wood I could find. My foot ached. Most annoying was the fact that the afternoon was nearly gone and I had accomplished nothing worth while. When one top crust turned out to look like the silhouette of an Indian, complete with feather headdress, I disgustedly gave the offending rolling pin a push and finished the crusts by flattening them with the palm of my hand. The results were only fair; but still they were better.

When the apricot pies were ready for baking, I unlatched the heavy oven door and let it drop. The whole stove lunged at me, quivered, and settled back again—an opening appeared in the joint of the stovepipe that allowed a noiseless expulsion of velvet-black soot to settle gently over me, over the stove, the pies, the pastry board, the floor, and over part of the table set with plates and cups, ready for supper. This was the final straw. The only advantage of this horrible mess was that it revealed the secret of a camp cook's fluffy biscuits to me. He was forced to open the oven door gently (as all my cookbooks say is necessary) or risk tipping over the stove. Hence, perfect biscuits every time in *that* man's lumber camp.

Dejectedly and nearly in tears, I limped to the office for help. The situation was not as tragic to Howard as it seemed to me. He obligingly pushed the stovepipe together, and helped blow the soot from the pies and arrange them in the oven; then he dipped hot water into a pan from the steaming barrel beside the stove. While I soaked my aching foot, Howard brushed the soot and flour from my face; and I forgave him all. And while the pies baked, my logger put a pan of left-over potatoes on the stove to brown for supper, capably sliced cold meat and cheese, cut stacks of bread, made with continued ease and efficiency a pot of fresh coffee and one of tea, discovered a tremendous camp-sized cake in the bread box, hard-boiled dozens of eggs—heated a gallon of tomatoes and a panful of corn. Then, because he liked them very much himself, this wonderful cook set dishes containing pickled pigs' feet on the table already crowded with pickles, jam, peanut butter, doughnuts, and cookies. (Today, a difficult cooking project for him is making coffee in our *automatic* coffee-maker.)

Before Howard called the jacks to supper, we gently, *very gently,* lowered the oven door and with the pie rake pulled out the flattest, queerest-looking shapes that ever could be contained in a round pie tin. The shortening in some pies had hardly melted, while other pies were on the verge of being cinders. Only one of these things could be faintly recognized as a pie fit to eat. As I stared at the horrible pies, I knew who invented the by-guess-and-by-gosh method. To this day I do not know what Baldy gave the crew for Sunday dinner dessert, but he did not give them my pies. Why the jacks did not quit was also a mystery to me—that is if I kept on believing what Baldy told me about not having pie for Sunday. Without ever mentioning the dismal failure my pies were, and without teasing me a bit about them later—as he jokes about my cooking failures now—Howard that same Saturday night hired a new cook in Hurley.

"Hang On to What?"

April twenty-ninth, the Saturday after I tried to bake pies, Howard and I were married in an Ontonagon parish by a minister as Cornish as a pasty and as plump. We decided suddenly not to wait until school was out in May to be married, but to get married and keep it a secret so that I could teach for the remainder of the school year.

The morning of our wedding day we drove to Ironwood, where I hunted for a wedding dress and accessories, and Howard searched for gardenias, which he knew were my favorite corsage flowers. The fragrance and loveliness of those he found were as sweet and beautiful as the words of the simple ceremony that united Howard and me. Because Howard could not find a platinum band to match my engagement diamond which was small enough, we were married with a borrowed ring three sizes too large.

Early in the evening Howard and I drove from Wakefield to Ontonagon on what we hoped our friends would think was a Saturday dinner date. Howard's aunt and uncle from Ewen, who had made the arrangements for us to be married there, met us at the parsonage.

As we took our places before the minister Howard noticed that I was trembling.

"You're shaking, Julie."

"I know it. I'm scared."

"Here, hang on to my hand."

—And suddenly we were married. For some unaccountable reason, the ceremony seemed shorter than the minister's impromptu advice and blessing. At last we escaped—Mr. and Mrs. Anderson—and drove to Ewen, where we spent the week end with Howard's aunt and uncle.

We postponed our wedding trip until later. Monday morning we returned to Wakefield. We told no one of our marriage, and I continued to teach. Howard was busy getting the last of the logs at his dad's camp skidded to a landing, where they were loaded on flatcars. At the same time, he directed the loading and hauling of his own logs on trucks.

After the last logs at Louie Anderson's camp were loaded on cars and sent to the paper mills, the lumberjacks strapped their packsacks on their backs and went down the road. Before the old logging road became impassable, all of Hungry Louie's camp equipment was loaded on drays, hauled to the highway, transferred to trucks, and taken to Howard's camp. Several weeks later, during breakup, it became impossible for Howard to log because of the spring thaws and because the jacks were quenching their perpetual thirsts.

"Do you want to go with me to Dad's camp to pick up the things left in the office, Julie? Instead of hiking over the old logging road through the timber, like we did last fall, we'll go in on the railroad spur and use the speeder for bringing the van and other stuff to the road where we can load them into the car."

"Sure, Howard, I'll go along. I've never ridden on a speeder. That'll be fun."

We put a can of gas in the car and drove to the railroad spur, where Howard tugged and pushed, with strength he did not appear to possess, until he managed to get the speeder out of hiding in the slashings and onto the tracks. After filling the tank with gas, Howard showed me where to sit on the speeder and told me, "Hang on. We'll be ready to go in a minute."

"Hang on to *what?*" All the speeder offered were a few

planks wired insecurely to something underneath the platform they made. A plank banged me on the forehead when I climbed on one end. I tested them all, and not one was fastened at both ends. Howard was busy getting the motor started; and, when he did, the noise it made was so loud that I never heard what he said to hang on to. I sat down with my legs crossed Indian fashion and grabbed a metal handle protruding through the insecure platform.

"Don't touch that. That's the brake," Howard shouted above the noise of the speeder. "Here, hold this." He pushed the gas can in front of me. (I dislike the smell of gas.) Howard gave the speeder a push. When it started moving down the tracks, he jumped on beside me. The contraption rapidly gained speed, and we were racing down the railroad. The pale yellow and green foliage became blurred walls of color confining my view to the rusty converging rails.

"Do you have to go this fast?" I screamed.

"Only one speed on this motor, and I guess the brake isn't working."

Trying to be casual, I asked, "Does this haywire thing ever leave the tracks?"

Howard did not answer me. He was busy trying to make the brake work and apparently enjoying the trip. Now, I realized he was intent on giving me a ride I would never forget. And I have not. At the moment I asked about going off the tracks, we hit a wide break in the rails and headed for the brush. I shut my eyes, expecting to land in a patch of berry bushes. We did—by stages. I arrived first, head first; Howard next, feet first; and the speeder last, upside down. The gas can was right side up along the track and still contained gas. We pushed and pulled the speeder; we grunted and moved it up the bank by inches onto the rails. That arduous task accomplished, Howard started the motor again, gave the speeder a push, and we were riding faster and faster down the rails again. I was amazed that the speeder had not been damaged, and shouted, "How can this rattletrap thing stand the side trips without getting all banged to pieces?"

"Oh, this old speeder is just like a drunken lumberjack, so relaxed all the time that nothing serious could happen to it. There's lots of give to this old thing."

A loose board banged me on the shin, and I agreed. "It surely does have lots and lots of give." I felt the first bump on my forehead and secretly wished the speeder wouldn't act so drunken.

The rusty rails curved into the timber. Howard called to me above the motor's noise, "Drag your feet. We're going too fast to make this curve."

I did the best I could to stop us, but the speeder continued on at a rapid speed. My feet bounded from one tie to another. Sometimes they caught on loose ties and I was nearly dragged off the platform to which I tenaciously clung. When Howard discovered that we could not slow down enough to make the curve, he shouted, "Jump!"

Never having jumped from much else except a springboard, I held my nose with one hand, put the other up above my head, shut my eyes, and jumped! Needless to say, the roadbed was a far less comfortable place to land than the old swimming hole. Howard slid off the speeder carefully and ran a few steps beside it. He slowly came to a stop standing up. While I dug the gravel out of my knees and hands and listened to Howard tell me how I should have jumped from the speeder, we watched it careen down the tracks, bounce along the uneven rails, and—to our complete surprise and dismay—ride merrily around the curve and out of sight without us!

We gasped! "Well, what do you think of that?" asked Howard. "Without us she rides the rails like a pullman. Guess we'll have to walk to the landing."

I was glad to be rid of the loose-jointed contraption, although I could feel that walking was going to be painful until I recovered from the soreness of my bruises and scratches.

Not far beyond the curve we came upon the speeder—off the tracks! Howard coaxed it into position for another flight into the timber. On the next excursion over the bank, we all rode together, stopping abruptly beside a wet, marshy, place. I thought we had been whisked away to a fairyland. From beneath their deep green foliage, royal purple violet-faces peered. We struggled with the speeder again and continued our ride to Ethelwood Landing.

The grounds there showed signs of previous loading activity, but all was quiet now. The only sounds to take the place of men's voices and the jingling of horses' rigging were a woodpecker's hammering as he circled a rampike searching for bugs and the whistle of a pine grosback as he called his newly acquired mate. From the office we carried bedding, a three-legged stool Howard's dad made years before from a crotched tree, a box full of articles from the van counter, and Louie Anderson's bookkeeping ledgers, all of which we piled on the speeder.

There was hardly room for us to ride. I weighted myself down with the ledgers and wrapped my arms around the stool, hoping not to take any more flying dives through the air. (It would have taken a ton of lead to keep *that* speeder on the tracks.) On the first side trip over the bank, I landed sitting in the brush on the three-legged stool wrapped in a blanket. During the second trip into the timber, Howard disappeared. I tracked him by following scattered mitts and socks. When I found him, he was struggling to remove the empty box from his head and shoulders. He told me later that it was a good thing what he said was muffled by the box.

After each excursion, we piled everything we could find back on the platform. Each time there was less. Howard's only comment to me was, "We're lucky. Once I took a case of eggs out to camp for Dad on this speeder. When I finally got to camp I had an egg case."

What we salvaged that day did not seem worth all the trouble. But, of course, we might have come out with everything *if* the brake had been working and *if* the tracks had been repaired. We *pushed* the speeder into the concealing brush (the one time we wished it would go there of its own accord) to hide it until it would be needed again. As far as I know, the speeder is still there, deep in weeds and bushes. We have never used it since that day.

Later during breakup season, before the woods roads had dried, we spent our honeymoon laughing about our excursions on the speeder and dining and dancing at the Radison Hotel in Minneapolis. However, because it was an introduction to the unexpected events which made life as a logger's wife difficult as well as pleasant, I consider the ride on the speeder our real wedding trip.

Marrying a logger has bound me to this rugged country, for Howard is as much a part of this region as are the giant pines and hardwoods. Every year my transplanted roots grow deeper and deeper, nourished not only by my husband's love and indulgence, but by the song of the wind in the tall pines, the flaming autumn beauty of ancient birch and maple trees, and the dazzling whiteness of winter.

But there were times during the first year after Howard and I were married when I would have traded all the timber on the Upper Peninsula and thrown in all the mosquitoes, flies, mud, manure, and plug tobacco at the Cedar Camp for good measure in exchange for one Chicago cinder in my eye. The pungent smell of cedar and pine only made me moan for a single whiff of Chicago's stockyards. Even the smell of Higgins ink or the sight of my pens and brushes gave me that nostalgic feeling one gets when reminded of a lost love. There were days when going to Ironwood nearby, to shop for absolutely unessential articles, was absolutely essential for my remaining reasonably happy. To stay up here, two Paul Bunyan jumps from the wilderness and become mosquito-bitten, mud-caked, blistered, and, finally, a faithful wife, when I would have preferred being a struggling fashion artist in cosmopolitan Chicago was not easy.

"Loggin's a Gamble"

In spite of breaking his arm, having his cook quit in the middle of the winter logging season, having to make new contracts to sell his dad's logs, and also traveling many miles to oversee two distantly situated camps, winter logging was profitable for Howard, and he was able to settle his father's and his own accounts. Shortly after we were married, he began looking for a stand of timber suitable for summer logging.

Near Lake Superior along the Black River and not far from his own winter camp, Howard found a dense stand of cedar in a swamp surrounding the camp of a Finnish logger who had used the buildings the preceding winter. This logger had cut the log timber but was forced to leave the cedar tie timber standing because the summer had not been hot enough to dry the swamp nor the winter cold enough to keep the roads continually frozen and make the cedar accessible. Gambling on the summer weather—hoping that the entire season would be hot and dry—Howard bought the cedar swamp timber at an extremely reasonable price.

We rented a small flat in Wakefield. For several weeks after I finished teaching, we were busy buying simple furnishings. One thing I had to find room for was the stack of ledgers which survived the speeder trip. Curious, I opened their big, heavy covers. Their yellow pages told a story of profitable logging when pine stumpage was available at low prices, when a cookshanty could be operated at small expense, and lumberjack labor was cheap.

My days alone in the apartment were long. With nothing urgent to occupy my time, I read with interest this record of things for which a logger spent his fortune. The entries recorded purchases Louis Anderson had made—logging equipment and supplies which changed with the times; horses, sleds, drays, and infrequent purchases of large quantities of dried prunes and apricots, salt pork, and cured meats; then engines, and Russell cars; and later, tractors, trucks, and frequent purchases of fresh fruits and meats.

Vividly the pages recording small payrolls and few purchases showed Louis Anderson's struggle through the lean years when his only operations were cutting and shipping the pulpwood from two forties east of Wakefield. Thumbing through the pages, I found want lists for the van counter and the cookshanty, and scale reports, records of sawyers' strip counts, and current prices he had been paid for pine and hardwood logs and pulpwood. Among bulging envelopes of canceled checks, I found one made out to Howard for seventy cents. Howard said, when I asked him, "That check, let's see. . . . I needed some spending money for the Fourth, I guess. Dad paid me that check after I had piled four cords of wood for the cookshanty stoves."

There were also records of the weather, prices Louis Anderson had paid his lumberjacks for cutting, cruisers' estimates on many sections in the Upper Peninsula, stumpage prices, shackers' accounts, a notation not to forget Isabel's birthday, and on the back pages of one big volume were entered Michigan Unemployment records. Obviously Howard's father had put the headaches of a logger's bookkeeping in the back of his books in order not to encounter them often enough to make logging unpleasant. A year or so later I was to understand his procedure and sympathize.

I stacked the bulky ledgers in a corner of our tiny closet, where they gathered dust until one memorable night when, without any warning whatsoever, the slats of our bed broke and the bed collapsed. The headboard smashed down upon us, making us surprised sandwich filling. In a dazed condition, we crawled from under the headboard to view the damages. One slat was splinters and another had come unscrewed. I applied a cold cloth to a lump on Howard's forehead, then I looked around for something to use for a slat.

When I found the ledgers stacked in the closet, I suggested, "Let's prop the springs and mattress up with these thick ledgers." It was a good idea which did not disturb the other occupants of the house and gave me more badly needed closet space. We worked until we had the bed propped up on the ledgers, then went back to sleep.

Several months later a man carrying a bulging briefcase appeared at our door and asked to check Louis Anderson's unemployment records. Quickly I recalled the struggle we had in piling the books under our broken bed.

"Oh, but you can't. Not now. It's out of the question. Can you come back when Howard is here?" (I thought of the dust under the bed.)

"Well, that's strange. If the books are here, why not let me see them? If there's anything wrong in the records it won't help not to let me audit them." He produced his card, which proved him to be an auditor for Michigan's department of unemployment. He eyed me with suspicion.

All I could say was, "Please come back when Howard is here."

"But you don't understand, Mrs. Anderson. You'll be saving me an entire afternoon if you let me see the books now."

I decided to tell him the facts. "The books you want are wedged tightly under our bed, holding up the mattress and springs. If you feel up to the work it will take to remove them, you may see them. If you come back this evening, you'll be spared an entire afternoon of struggling." As I expected, he left in a good humor and returned that evening when Howard was home to do the *menial* book work.

While the crew built plank roads into the cedar swamp, Howard drove back and forth every day to camp from our apartment in Wakefield. Because the thirty miles to camp each day consumed precious time, as soon as the roads were finished, the bunkhouse filled with sawyers, and a cook hired, Howard, too, stayed at camp during the week, leaving me alone from Monday until Saturday, which I didn't like at all. The office was small, but large enough for a barrel stove, a bed, and a counter for supplies the lumberjacks would need. With nothing to keep me in our apartment, I began staying with Howard in the office during the week, too. After the jacks had been paid on Saturdays, we would come back to our apartment in Wakefield to cool off and clean up.

The van we salvaged from our wild speeder ride Howard took to the new Cedar Camp. I stacked it neatly under the counter in the office so that Howard could find what the jacks wanted in a hurry. Always, just after I had straightened everything, one request for a single article in plain sight left the neat piles a mess. If there was nothing to read or no one to talk to, I could always straighten the van counter.

"How did this stuff get such a queer name?" I asked as I piled the lumberjack shirts and woolen underwear in neat piles under the counter.

"Van," Howard explained to me, "is anything a logger has for sale in the office to satisfy the wants of lumberjacks. In the old days, the logging camps in Michigan were visited by peddlers bringing vans stocked with many articles and a large variety of clothing. The lumberjacks called this wagonload of articles 'the Van'—hence the name 'van' today. Most loggers' time sheets, like mine, have a column headed 'van' which includes, mainly, tobacco, soap, candy, clothing, medicine, and chances on the punch board."

Such constant labors, to keep things going in any man's logging camp! If the van counter at the Cedar Camp was in order, there was always the office floor to sweep. The unpainted plank floor never stayed clean a minute unless I hooked the latch so no one could come in. During the rainy spring weather, the jacks tracked in mud and manure; in the summer they brought in manure; during the fall, mud and manure; and in the winter, snow and manure. As fast as they tracked the floor, I swept it. As soon as I heard the bunkhouse door slam and saw a jack start for the office, I grabbed the broom. I woke up at night to sweep the floor. Howard said, "I think you wear out brooms just so you can use the handles for curtain poles."

Our continual buying of camp brooms was an expense only slightly less than the expense of buying fly spray. The monotone buzzing of flies was as unbearable to me as the incessant blowing of the prairie wind had been to my mother 'way back in Nebraska. The office windows were always hazy with fly specks in summer and hazy with smoke in the winter. Idleness at camp was sheer laziness, because there always was manure, flies, or smoke.

The road from camp to the highway was graveled except for a stretch of plank road over a part of the swamp which had been logged. At the roots of the stumps and rampikes there were pools of murky, slimy water. Even though the summer was hot and dry, these pools never completely dried. They became the perfect breeding places for vicious vampire mosquitoes. The buzzing of the mosquitoes was deafening. When one mosquito swooped down to attack its victim, a noise like that of a dive bomber could be heard over several sections. When six or more dove after choice morsels, the sound was worse than a celebration of Navy Day. We could have endured the noise if their bites had not been so crippling.

One Saturday evening on our way in to our apartment in Wakefield, where we spent the week ends, our car skidded off the slippery planks. The rear wheels sank down into the mud up to the axle. Howard found a pole with which we tried to pry and push the car back onto the planks. Just as we felt the car rising a little, along zoomed a dozen mosquitoes, bloodthirsty monsters, to tipple a new liquor. We swatted and swatted and swore; then the car sank into the sucking mud again. Bitten and bleeding, I made a suggestion. "Howard, I'm going back to camp and get the big black kettle the jacks use to 'boil up' and a length of the blacksmith's heaviest chain. I want revenge."

Howard scornfully interrupted. "How in the devil do you think you're going to get *revenge* with a *kettle* and a *chain*? I'd suggest usin' arsenic as bait in a bear trap," he growled ruefully.

"Well, I'd chain the kettle to that big maple over there, crawl under the kettle as bait—being sweeter meat than you—and let the mosquitoes drill through the iron kettle with their beaks. Then I'd bend their stingers on the underside of the kettle. That way I'd hold them, at least until you get the car back on the planks."

But foresighted Howard came right back with, "That would be wasting your work. These mosquitoes are so darn big and would strain so hard to get free from the kettle they'd uproot the tree. They'd carry away the kettle, chain, and all. The draggin' tree would gouge out a helluva ravine; we'd have to build several bridges over it, so we could get our cedar out. That would cost so much it would use up all the profits I hope to make this summer loggin' the cedar swamp."

I gave up my plan and resumed swatting the beasties, squashing their gorged bodies on my neck, face, and arms; sacrificing my good red blood for a profitable summer of logging. We spent an hour trying to pry the car back onto the slippery planks. Howard finally gave up and went down the road half a mile to borrow a skidding horse from a Finn woodcutter. If the horse had slipped into the mud, too, I believe I would have gone back to camp and stayed under the black kettle for the rest of the summer. But in a comparatively short time the horse pulled the car back on the planks; it was days, however, before the swellings from the bites went down and my face, arms, and legs became their normal size. Howard's long-sleeved shirt, his pants tucked into high boots, and the stubble of red whiskers on his face protected him from the mosquitoes. I was exposed to them all summer.

And then the termites. During the nights we stayed at the Cedar Camp, I was kept awake by their continuous chewing in the boards of which the office was built. I was lucky enough sometimes, however, to fall asleep before they started their all-night gorging. The office did not cave in; but, because of the termites' greed, I lost many hours of sleep listening and wondering where the next hungry worm would begin his chewing. They seemed to find the wood closest to me the most tasty. When the termites and the heat kept Howard awake too, we swatted mosquitoes. And, through the doorway, we watched the moon appear above the trees.

"You know, Julie," said Howard, one such night, "loggin' is a big gamble—has been ever since pine timber began to get scarce in Lower Michigan. There are big mistakes a logger can make when he buys a hard loggin' chance or pays too much for his timber. Risks like that make him lose his shirt, but it's the uncontrollable things he gambles on to make a profit. Loggin' that pays off is governed to a great extent by the weather. If it rains much and stays wet this summer, our chances to get all the cedar out of the swamp are slim. The less timber we can get out, the smaller the profit'll be.

"In winter loggin', an early freezin' cold night might catch a logger with all the radiators in his trucks and tractors still filled with water. It takes time and money to replace cracked radiators. When snow fills the roads, there's your all-night plowin'! That added cost cuts down the profit, too. And, too, the deep snow covers logs we don't find until too late to skid them out in the spring—if ever. Unreasonable cold weather makes tractors and jammers hard to start. And, occasionally, the fires built under the motors to warm them heat them so fast that parts crack and cause future bigger delays.

"And a logger dreads fire that sweeps through dry timber after a hot dry summer and fall. Fire's a thief—it steals timber, equipment, and camps and it never makes reparations of any kind. If breakup comes unexpectedly early, there are sure to be logs that cannot be reached because they are along roads deep in mud. The rain that falls day after day some springs and keeps the jacks in the bunkhouse is just trouble to loggers; the cuttin' and haulin', of course . . ."

Just then a rattling good car skidded into the camp yard. Two jacks got out. Loudly they decided they were in the wrong camp and climbed back in the car. The driver ground the starter for fifteen minutes and would have kept that up all night, I believe, if Howard had not gotten out of bed to start it for him. On his way out of the office Howard let in a new batch of mosquitoes—and another swarm on his way in. We began our swatting all over again.

Howard and the mosquitoes droned on. "Rain can even more remotely cause costs to get out of control. During depression days Dad cut short pieces for one of the mines. He sawed and hauled the mining timber for all he was worth. The check he would have got for it wouldn't have been large, but in those hard times it would have been very welcome. When Dad delivered the mining timber the foreman thought they didn't look to be the right length. He drew his rule and measured them. 'My gosh, Louie,' the foreman said, 'these pieces of mining timber are six inches too short, accordin' to my rule!'

"Dad was dumfounded. He got out the tape he had used when he cut the timber; to his dismay he found that the tape measured six inches short. Dad decided that the tape must have shrunk from having been left in the rain at one time or another. He bought a new steel rule and went back to the woods to cut and haul more mining timber, measured with a rule that couldn't shrink."

"I've heard enough tough-luck stories about logging. Isn't there something encouraging about the logging game?" I was sure now that tomorrow it would begin to rain and keep on raining every day all summer—just because we were depending on dry weather to get the cedar swamp cut.

"Well, most loggers help each other out of tough spots if

they can; they lend each other equipment and they use each others' roads and landings. A pretty good lot—these loggers."

"Of this you can be sure, Howard, you're my logger and I'll stick by you through thick mud or thin, big mosquitoes or little no-see-ums, deep snow or none, rain or shine."

Howard drew me close to him. I had married a logger and now I chose to be miserable living his way with him rather than walk out alone into a life of familiar and comfortable things. This most important deal of our lives we closed with a kiss. The honey-colored moon slid behind a cloud. The noises of the warm summer night settled into a deep, enveloping hum that carried us off to sleep.

Most summer nights at camp after I had gone to sleep, I did not wake up until the sun tinted the horizon with shades of cerise. But one still night I woke up and decided to make a trip down our private path. I slid my feet into Howard's handy house slippers and clopped out of the office and down the path. Full moonlight gave the camp buildings and the trees a frosty look. A tiny breeze made the leaves shimmer and gleam in the silver light. Thre was not a sound except the scraping of the big slippers dragging on the soft dirt of the path.

I had gone only a few yards when I looked ahead and directly at the black bulk of a bear. He clutched in his huge paw butter the cook had left to cool in the spring behind the cookshanty. The bear made no advances. He blocked the path and growled deep and low. I supressed a scream! The bear growled louder. He shifted his position to come toward me. An owl perched directly above me on an overhanging branch hooted—wise old owl—scaring the wits still further out of me. Still growling and clutching the yellow butter, the bear ambled off into the brush while I made a split-second rightabout-face and ran to the office. I got there barefooted and shaking like a leaf. I screamed in Howard's ear, "There's a bear behind the office stealing the butter."

"Bear stealin' butter, you say? I'll have to tell the cook

to hide the butter better." He mumbled something about butter being quite high-priced and began snoring.

I crawled back in bed, shivering from fright and the night dampness. I wondered which had frightened me more, the bear or the owl. Not until the sun was shining did I go back to look for Howard's slippers. They were both under the branch on which the owl had perched; they were headed down the path. . . .

The heat that summer was intense and continual—perfect weather for logging in a cedar swamp. There was little rain, and swampy places which were never known to dry became downright dusty. The flies were as carnivorous during the day as the millions of mosquitoes during the nights. Worst of all the pests were the tiny black flies, aptly named "no-see-ums" by the Indians; these insects can hardly be seen, but certainly bite in the most difficult places to scratch. The heat and flies drove me to the cookshanty for coffee. The heat of the cookstove drove me to the woods where it was cooler. But the pests were there in droves—much worse than in the open places. The black flies drove me back to camp.

When I could coax Tex away from a shady dog's heaven under a cedar, he went with me to Little Girl Point on Lake Superior only a few miles from camp, where I cooled off by going for a swim. Tex rode outside the car on the fender with his front paws wrapped around the back of the headlight. All I could see of him through the windshield was his head pressed close to his paws. The wind slicked his fur close to his head and occasionally bent his ears backwards. The faster I drove, the more he enjoyed the ride. Usually he would have nothing to do with the lake shore. After one sniff (I could see him quiver when the smell of dead fish hit his sensitive nostrils), he would go back to the shade of the car to wait until I had enough of Lake Superior's icy cold, white-capped waves.

Range people, by retelling, have kept alive a romantic Indian legend about Little Girl Point that was first told by a certain Henry Schoolcraft a century ago. He learned the story from his wife, a Chippewa woman of mixed blood, who belonged to the Waubojeeg family that lived along Chequamegon Bay during the Revolutionary War period.

The legend concerns Lilianau. According to Henry Schoolcraft's story, this girl lived at the foot of the Kaug Wudjoo (the Crouching Porcupines). A husband had been chosen for her by her parents. The day of her wedding had been set.

As the day of her wedding approached, Lilianau accompanied her father on a hunting trip to the top of the Porcupine Mountain range. He warned her not to cross the Sacred Grove that extends westward from the Kaug Wudjoo along Gitche Gumee's shores to the point now known as Little Girl Point. Like many others who had been warned, Lilianau entered the woods, disappeared, and never returned to her father's lodge at the foot of Kaug Wudjoo. Her people thought she had been carried away by the Puk Wudjies, the fairies of the woods.

Very few of Lilianau's people invaded the region of the Sacred Grove. The trails they followed between Kaug Wudjoo and places along Gitche Gumee farther west were by water. One starlit evening a party of Indian hunters were returning to their village at the mouth of the Ontonagon River. As they passed the point west of the Sacred Grove, they saw in the dusk a female form beyond the trees along the shore. Paddling cautiously nearer, for they were afraid to land, they recognized the face of the lost girl of the Kaugs. She was dressed in green. A youth who wore a green plume in his hair accompanied her.

Some folks say that Lilianau had fallen under the enchantment of the Puk Wudjies. Others say that the youth wearing the green plume was her lover, one of her own race with whom she had kept a tryst.

Thus the place where the lost Lilianau was last seen became known to the Indians as "Point Lilianau" and to the early French explorers as "Le Point Petite Fille." The pretty

French and Indian names have been discarded, and the point is now known as "Little Girl Point."

As I relaxed on the warm, white sand of Little Girl Point I mused about Lilianau, who I chose to think kept a rendezvous with her lover. The entire point became imbued with romance for me.

Few people frequented the point during the week. Many times I was the only person there. Then Tex walked along the beach with me, resting when I did in the cool shade of a bleached pine log that probably had floated in from a boom being towed to Ashland when the lumberman John Schroeder logged along Superior's shores. The rugged shore line of rocks, white sand, smooth driftwood, and scraggly pines and low bushes took tolerantly the rhythmic lashing of the cool green waves. This natural acquiescence in nature was restful to me, because at the Cedar Camp our struggle against the elements was continual.

Refreshed by our escape from the pests and heat at camp, Tex and I would go back late in the afternoon when the jacks were coming in from the woods, their chambray shirts soaked with sweat and their hair matted to their foreheads. The horses' shiny coats were splotched with dark patches of sweat. The beasts drank eagerly at the watering trough fed by a spring. Lumberjacks would be gathered around the pump gulping down the cold sweet water and splashing it on their hot flushed faces. Other jacks were slumped on the bench in the shade of the bunkhouse to rest before cleaning up for supper.

In spite of all the pests, heat, and mud at the Cedar Camp, by early fall the jacks had produced over seventeen thousand ties, many cords of lagging, a large amount of pulpwood, and some sawmill logs. What he made from these products Howard invested in a section of timber south of Tula along the Presque Isle River. At the same time he got an option on a piece of timber north of Tula across the highway. As timber stumpage prices steadily increase and large tracts available decrease, loggers have learned to buy small stands of timber that can be cut quickly, which enables them to realize profits on their investment in a comparatively short time. Then they let the property revert to state ownership, which relieves them of paying taxes on cut-over land. Since the option on the timber gave Howard first chance to buy that stand across the highway, he had obtained enough timber potentially to keep him in the logging business for several years.

"How do you know you want to be a logger all your life?" I asked Howard, hoping that there might be something else we could do summers rather than be mosquito bait.

"There is this about loggin', Julie, that you don't understand—*yet.* It gets into the blood like the yen to gamble."

"Via mosquitoes?"

Howard gave me a scorching look. "The more a man logs, the more he wants to log. No amount of losin' keeps a real gambler from bettin' his last dollar on what he hopes will be a winner. No amount of misfortune keeps a real logger from makin' another try for the fortune he hopes to make loggin'. Dad wanted me to be a lawyer, and I might have if I hadn't found the excitement in loggin' camps far more fun than studyin' can ever be. We may have tough sleddin' some years; but loggin' was the game for Dad, and loggin' is the game for me, too."

I didn't argue. His courage made me love him more. "I doubt if our luck of this summer will hold," was all I said. I had visions of our small fortune being washed away the next summer by incessant rains; if not that, then consumed by fire or lost in snowdrifts. Howard retained his optimism.

While Howard was planning the road into the new camp near Tula, I tramped along with him through the timber. We made our way along a trail he had blazed as the possible future road to the camp. The leaves of the tall maple and birch made a ceiling of palest yellow and brightest flame, with patches of blue sky showing now and then. There was very little underbrush.

Early fallen leaves and broken branches and twigs carpeted the soft springy ground from which there rose a fragrant warmth like an incense. Together we walked where now there is a gravel road; then our path wound up inclines, down ravines, and around stumps of fallen trees and moss-covered logs, with shafts of warm sunlight shining through the trees and falling occasionally across our way.

When we came to one high piece of ground, Howard stopped. "This looks like an ideal spot for a set of camp buildings. It's not centrally located on the section, but I think we can find water easily here. The only spring is far back on the south side of the section and couldn't possibly supply us with all the water we'll need for both men and horses." He pointed out the general location of the new cookshanty and bunkhouse. "On this knoll we'll build the office facing the bunkhouse. On the south side of the office there'll be a lean-to that we can use for a bedroom when we stay at camp. That'll be better than sleeping as we did this summer—right in the office."

"We won't give up the apartment, will we?" I hopefully asked. After enduring the bareness of unpainted floors, curtainless windows, camp benches, and a lumpy bunk bed to be with Howard much of the summer, I realized what a heaven our town quarters were.

"Certainly not," Howard assured me. "The apartment is home to us. The office in a camp is where a logger stays while he is on the job loggin', comparable to the hotel room of a salesman. Saturdays, most loggers come home to play with their children, take a bath, take their wives to a show; and then on Sunday they want to eat a so-good dinner and fall asleep listenin' to the radio." Teasingly he added, "Where in the devil would I take a bath if we didn't have a home? In the creek?"

"Frankly, I don't know, but I am very glad to be keeping the apartment," I answered.

As we walked back to the car through our section of virgin hardwood, the sun warmed our shoulders when it could shine down through the treetops. We left behind the untouched forest which would presently be cut by a maze of logging roads and skidding trails. The ring of the ax, the noise of the saw, the jingle of rigging, and the voices of men would take the place of bird and animal noises. Log decks would grow high; and, finally, log trucks would pull out of the woods with the crop of timber that stood so proudly that day making dense shade in the autumn sun.

Several weeks later the leaves were falling fast. Sunshine flooded the forest. Showers of bright leaves covered the tractor driver as he pushed stumps and small hardwood trees, roots and all, out of the roadway. Sawyers ahead were cutting the big trees. Trees not yet skidded to the side lay across the roadbed with great gashes cut through their trunks. While I watched the road-building process, Howard came out of the woods. "I just found several hollow basswood trees to use as culverts and another that will make a good horse trough."

Station by station, men built the road to Tula camp so that trucks on this well-contrived system could wheel out of the woods in the winter with no treacherous hills or sharp curves. At the Cedar Camp and Howard's winter camp north of Ironwood, the crew was taking down bunks and packing all the equipment and supplies on trucks. They were getting ready to haul them to the new camp site as soon as the road into the timber was completed.

When the last truckload of camp equipment left the Cedar Camp, Tex climbed onto the fender of the car. Without looking back, Howard and I left the cedar swamp. A forest of hardwood on high land waited for us at Tula Camp. Ringing out already there sounded the cries of "Tim-ber-r-r!"

Part Two

"Tula Tullala"

Tula once thrived as a lumbering community composed of a shingle mill, a lumber mill, a general store, a blacksmith shop, barns, a long bunkhouse and separate cookshanty for single millworkers, and small houses for married millworkers. Today it exists only as one of the many now-deserted mill towns of the Upper Peninsula. The few buildings still standing are vacant and windowless, dilapidated and weathered. The South Shore trains passing by shake the skeleton buildings, and ghosts of the past are aroused from their sleep only to fall back into their lethargy after the train has gone by. I picked an armful of white ox-eye daisies along the road there one sunny morning not long ago. The daisies and Indian-red paint brush that grow there profusely are all the color left to that once-colorful community.

In the middle eighties, Tula was no more than a section house for the construction crews who laid the rails for what was to become, years later, the Duluth, South Shore and Atlantic Railroad. With the building of the railroad across the Upper Peninsula, the forest filled with the sounds of crashing trees and the ring of axes. Among the men felling the timber and laying the rails were stalwart Finnish woodcutters who sang the folksongs of their native land while they worked. Their harmonizing voices could be heard by the gangs laying the rails far behind them, and the haunting minor chords mingled with the ring of axes and the crash of trees.

At night, around their open fire, they sang their Finnish songs to amuse the other men and to make the lonely evenings pass more quickly. According to the story told to me by August Bye—a neighbor of ours who for six years lived at Tula and worked for the John Schroeder Lumber Company as bookkeeper and storekeeper—one delightful lilt, "Tula tullalla," became the favorite of most of the lonely men. Soon they were all humming:

Sophia Tuomi, our Finnish housekeeper often sings this charming folksong to our three boys. She says that a free translation of "Tula tullalla" would be:

Chubby-cheeked, chubby-cheeked,
Is the sweetheart of this boy.
Come to us, we are baking now,
You might get a piece of cake, perhaps.
Come through the door, don't sneak around,
And don't keep staring at me!

Out of their mutual liking for this quaint Finnish folksong, the men who cut the right-of-way and laid the railroad chose to call their section house and the immediate vicinity "Tula." And Tula it is today, even in its decadence.

After the men of that first section crew had named Tula, the South Side Lumber Company acquired the timber and set up a lumber mill and a shingle mill at Tula. Without even beginning operations, this lumber company promptly sold to the Tula Lumber Company, who engaged Fred Gregory to operate the shingle mill. Mr. Gregory, with whom I last visited just a year ago in Ontonagon, is an aging man with brown and weathered skin, eyes that twinkle, and a vivid memory. He told me reminiscently that the endeavor was a losing proposition from the beginning, but that the big catastrophe occurred during the following summer: fire rolled in from the slashings south of Tula! In spite of the tireless efforts of the mill men and lumberjacks to control the widespread flames, the shingle mill, piles of cut cedar logs, and some of the standing timber were completely destroyed.

TULA TULLALLA

The next year, 1913 (the year Howard was born in Ewen), with the intentions of logging the hardwood and making a mill at Tula a profitable enterprise, Lewis Jenson purchased the remaining timber along with what was left of the logging equipment and the buildings at Tula. With the help of Howard's dad, he supervised the rebuilding of the mill and made roads deep into the timber that summer.

In the fall, Howard's aunt Louise Goodenow came to Tula with her husband, whom Mr. Jenson employed as assistant superintendent of the mill. Mrs. Goodenow kept the records and accounts at the mill and also acted as postmistress. She and her husband lived comfortably in one of the small houses occupied by the families of the men who worked in the mill. Many sawyers, swampers, teamsters, and cooks came to work in the timber around Tula and lived in the logging camps miles from the railroad. The life of these lumberjacks was typical of life in other logging camps. They logged until they had a stake; then they "pulled the pin," cashed their time slip at the company store, and boarded the train to Ashland during her boom days. Or, they went farther down the line to Superior and Duluth. Some jacks preferred to ride the train only as far as Thomaston, walk south a few miles to Wakefield, and then go on to Hurley.

The community of Tula developed around its industries of logging and producing lumber products. But all the activity while Howard's Aunt Louise lived at Tula was not directly connected with logging. Most of the small houses, although they were not constructed too well, became pleasant and homelike quarters. Gay curtains, geranium plants, and knickknacks brightened the sparkling, small windows of many homes. During the day Aunt Louise watched happy children play in the snow, or she took time out from her bookkeeping to enjoy a cup of coffee in the kitchen of a friend whose home smelled of fresh-baked bread or of sudsy water and freshly washed and frozen clothes.

The millworkers' families spent some of the long winter

evenings visiting at their neighbors', singing, jigging, or story-telling. More often, tasks to be done kept them home. In the soft light of old-fashioned lamps, women pieced and stitched quilts while they discussed in low voices the day's events with the men, busy greasing boots or hanging socks to dry by the stove. Girls darned mittens and socks or patched under-wear. Mothers read stories to sleepy-eyed, rosy-cheeked children—the same stories, no doubt, that Howard's mother read to him and my mother read to me. Lovers courted in the shadows or pulled a batch of taffy and popped corn over the woodstove in the kitchen—simple pleasures giving more enjoyment than all our foolish search, these more recent days, for entertainment and diversion.

For a number of years (years during which my folks homesteaded in Nebraska and Canada, and I drew multitudes of frisky cottontails), hardwood logging and lumbering thrived under Jenson's supervision. In the mill yard there stood many piles of drying lumber. In the woods were hundreds of cedar posts and short pieces ready to be hauled to Tula. The cicadas proclaimed the dryness of the weather day after day with their rasping song. In the swamps the mud dried to hard-baked muck. In the mill yard the dust rose in puffs behind the heels of the millworkers. Slashings were bone-dry, and the fresh smell of the cut cedar boughs was gone. Even in the shade, Aunt Louise said, the heat became unbearable.

Sensing that their immediate danger was fire, the mill-workers and lumberjacks tried, by digging trenches and throwing up the fresh sod on all sides, to protect Tula and the cut timber from flames coming in from the woods to the south. But without any warning, early on a Sunday morning when many of the jacks were still on benders, the barn boss, on his way to do his chores, discovered a blaze at the edge of the lumber yard.

Running from house to house, he aroused as many men as he could. The fire spread rapidly and soon was burning in the piles of lumber by the mill. Some men beat the flames with blankets; some dug up the dirt and tried to cover the fast-spreading flames as they consumed one pile of lumber after another, burning closer and closer to the mill. Others tried to save the mill machinery, while excited mothers and children carried their precious belongings to a safe distance across the railroad tracks.

With only a few men and no fire-fighting equipment, the job was hopeless. Shouting, sweating, blackened men surveyed the crumbling lumber piles, the clouds of rosy-tinted smoke, and the blazing mill. They realized the battle they were fighting could not be won. In despair they, too, crossed the tracks, where the heat of the fire had crisped the weeds and the ferns and where large black ashes of their burning homes floated down upon them and their families.

The mill was completely burned. Lumber piles where the fire started were flattened. The fire had burned some of the homes and then spread to the cut timber in the woods around which there was the fire trench to prevent the fire from coming from the slashings. There the fire stopped, having destroyed the community of Tula except for the barn, several cottages, the boarding house, and the store building. The loss was tremendous, and Lewis Jenson decided against beginning operations at Tula again. He gradually disposed of his timber, selling it piece by piece. Profitable hardwood logging did not begin at Tula until John Schroeder acquired the property.

Between 1919 and 1925 (years during which Howard's father made his fortune logging at Moran and later lost it) the Schroeder Lumber Company logged steadily at Tula. Jim Gough, one of Schroeder's cooks, still carries on in the style of logging cookshanties. At an abandoned logging landing near Tomahawk, Wisconsin, Jim has a cookshanty where he serves camp-style meals. As much as one likes of whatever one likes can be had in surroundings that transplant one back to the good old logging days of oxen-drawn sledloads of logs, river drives, muzzle-loading bunks, kerosene lamps, barrel stoves, baked beans, and salt pork. We never fail to stop at Jim's Landing for one of his overwhelming meals whenever we are within twenty-five miles of his log cookshanty.

Between the highway and Jim's place there stands a grotesque conglomeration of logging rigging that white pine loggers used years ago. Towering above weather-beaten sleds and silver-gray jammers held together with rusty irons, a pair of wheels fourteen feet high stands blackly silhouetted against the surrounding logged-over woods. The first time I saw big wheels they were a vague contraption to me. Levers and such things are not my line. But today the big wheels standing in many Michigan courtyards are a symbol, emblematic of a fabulous era of logging which left Michigan stripped of her vast pine forests and unbelievable wealth and dotted with deserted logging centers such as Tula.

Each visit to Jim's Landing leaves us steeped in logging lore, awed at Jim's ability to prepare such good food, and, incidentally, reveals another logging story.

One such story concerned a bull cook called "Brokey," short for "Broken Ass," a name he acquired because he always went about doing his chores in a bent-over position and with an extremely painful expression on his face. He was a Frenchman with black, black hair, which he always kept covered with a tassled stocking cap—even when he slept. He was good at playing cards and was known to have won more than one big poker stake of cash in Hurley. Toward the end of the winter, when poker games at camp were on the cuff or played for small stakes of tobacco, Brokey still had the cash—where, no one knew.

Once, while a poker game was going on in the office, Brokey came in with his broom and pan. After trying unsuccessfully to reach under stools and sweep around the boots, he exploded. "Jeez Christ, Jeez Christ! No can scrape-um floor. No can scrape-um floor. All time play poker. Never stop. All time play poker!"

One poker player jestfully grabbed Brokey's red stocking

Big Wheels — to Me a Symbol

cap and tossed it into the door of the stove, which Brokey had just opened. The greasy cap began to blaze. Brokey dropped his broom, grabbed his cap out of the fire, threw it onto the floor, and stamped the flames out excitedly! He let out a stream of cuss words; and, from the inside of the double folds of the cap, he pulled a thick wad of bills.

Brokey came to the office one morning scratching his head. Flicking lice into the fire, he asked for a sure cure for the "damned bugs." The foreman behind the high counter looked at his array of antidotes and medicines and handed him a bottle of hydrogen peroxide. "Here, Brokey," said the foreman. "This'll do the trick. Use it all, and you won't have to scratch."

Brokey used it all. With French fire in his eyes, he came to the office the next morning long before breakfast.

"Jeez Christ, Jeez Christ. I keel you, you Eenglish devil. I keel you." And he might have, if there had not been a high counter between the foreman and himself. In his rage, Brokey grabbed off his cap, to show the reason for his anger. His once-black hair gleamed fiery red. From then on until spring, the crew ribbed Brokey about his red hair. Brokey wore his cap all winter for two reasons, but mainly to keep the lumberjacks from reminding him of his unfortunate case of lice and his red hair.

As storekeeper at Tula, August Bye had his share of troubles, too. Not the least of his troubles was the problem of interpreting the needs of the lumberjacks who spoke little English. Many jacks could neither read nor write; and there were some who found it difficult to speak any language except their native tongue. One Swedish jack who came to Tula to get a job got off the train on a day when the weather had turned exceptionally cold. Shivering and shaking, he said, "Ay vant some kalsunder." He was offered hot coffee, a drink of whisky, and everything except what he wanted. Finally, the combined efforts of several other jacks brought out the fact that what he wanted was a pair of heavy woolen underdrawers.

One stormy day a tired and snow-covered Austrian sawyer came into the store at Tula for an "ipet decet snap." He had been lumberjacking in Michigan for five years and had, at last, written to his folks in Austria. He wanted a five-cent stamp to mail the letter.

The log cookshanty and the bunkhouse where Jim Gough baked beans and Brokey scraped-um floors are deserted sagging buildings in an open place in the slashings. Berry bushes, vines, weeds, and wild flowers are creeping in around the store at Tula where August Bye sold kalsunder and ipet decet snaps. "Tula tullalla" is a vaguely remembered tune. And colorful Tula exists only in retrospection. That day Howard and I left the Cedar Camp behind us and drove by on our way to our new camp south of Tula, a South Shore engine whistled mournfully as it approached the abandoned, silent town. As the train passed, giant puffs of smoke blotted from our vision Tula's vacant, gray buildings.

Willow Witchery

A short way down the road from Tula, we came to a cleared place on the south side of the highway where a gravel road turns and crosses the South Shore railroad at right angles. Howard drove onto this road and swung around a curve. And there, sprawled in a clearing, we saw an establishment he called "Watermelon Jim's Place." Staggering all over Watermelon's place were fences and shacks, barns and pigsties, chicken coops and garden spots, rubbish heaps and manure piles, a house and outhouse, and a garage and woodshed.

All the buildings were low, roughly put together tar-paper shacks except the house, which was well built in comparison. The shacks looked like weak-kneed structures; some leaned one way and were propped on that side, while others leaned another way and were likewise propped. Each year their roofs had evidently sagged more and more under the ponderous weight of heavy snows. The grotesque shadows of these outlandish buildings bulged and jogged because of a worn-out tire leaning against the garage, a ladder minus several rungs propped against the house, and an old oil stove—no longer usable—put out by the barn to rust. Odd lengths of lagging and sapling trees used as rails fenced off the barnyard areas. The fences gave evidence of repeated repairing but not enough to make them effective boundaries for the animals. Rows of vegetables and patches of potatoes filled the unfenced places between the buildings and the barnyard.

As we approached the place, a skinny black horse jogged to a distant corner of the barnyard. Mournful mooing sounded from the dark interior of the barn. Pigs, rooting in a sunny spot outside their sty, squealed louder. Chickens, scratching all over this maze, fluttered across the road in front of the car. We stopped before an iron gate which prevented us from driving through this conglomerate arrangement of buildings, fences, and gardens.

"This is the road I told you about that we can use if we keep the gate closed," Howard explained. "I don't know what good it will do with the rest of the fence so broken down, but Watermelon thinks this gate keeps his wayward animals from strayin' onto the highway and railroad. In return for the use of his road, I promised him he could haul the garbage from our cookshanty to his pigs when we start to log. I'll open the gate and you drive through."

As I started through the gate, I saw Watermelon running down from his house. He waved his arms and yelled in Italian. Around his middle his shirttail flapped. I almost smashed the car into the gatepost in my eagerness to stop and keep Watermelon pacified.

"What in the devil's wrong now?" I heard Howard mutter under his breath.

Scrambling downhill on an even keel was evidently a terrific exertion for Watermelon. The knot of the red handkerchief tied around his neck had slipped high against one ear. His straight, uncombed brown hair stuck out in every direction from under a ripped, hard straw hat and formed a fringe around his flushed face. Watermelon was fat and wore his belted pants below his bulging abdomen. In his excitement today (and being Italian, he was always more excited than calm, I found), he flung his arms about to supplement his statement. His shirttail, which had slipped from the confining embrace of his belt, emphasized his wild motions.

Watermelon arrived at the car puffing and red-faced. In one hand he clutched the cleanest, choicest bouquet of garden vegetables I had ever seen. These he dumped in my lap. "For Mamma" (meaning me) "to make-a dee-lish-us pot soup. And dees make it just-a so-so." He clucked his tongue as he added a handful of herbs from his mamma's herb garden. From his shirt pocket he drew a potato nearly skinless from scrubbing. I thanked him again and again, and was vastly relieved that we had not done something to make him angry.

Howard, just as pleased as I was that Watermelon remained friendly, kidded him about his beard, which was obviously a growth of several weeks. Watermelon stroked his chin, laughed heartily, and rolled his eyes. "When I shave off-a da whiskers, Mamma, she tink I make-a da eyes at somebody young."

I moved over to make room for Howard to climb behind the wheel, and we drove away. Watermelon shut the gate and waved to us awkwardly. Singing loudly in Italian, he climbed back up the hill. I found him to be an energetic, generous, and happy Italian, whose real name apparently had been forgotten since he acquired his nickname.

"Once," Howard told me, "to satisfy his liking for watermelon, Jim grew a patch of big watermelons. When they were perfectly ripe, he ate four whole melons. To eat that many melons would have killed anyone else—but not Watermelon. For three days he was too sick to talk intelligently. In his moanings and groanings the only word he uttered was 'water-melone—water-melone.' 'Water-melone' he is today."

Beyond Watermelon's gate the road wound around and up and down small hills. It passed by a pond and at one place was close to a gravel pit. The removal of gravel from the pit had caused the road to cave in on one side. A sapling stuck in the ground marked the sharply declining edge on the caving side. Up a steep incline from the pit, the road led across an open field—a solid patch of tall, waving grass.

"This field was once a potato patch, which Lewis Jenson planted and harvested to feed the crews of lumberjacks who worked for him when he logged around Tula," Howard told me. "It looks like this long stretch of road through this field is going to be hard to keep open when the snow drifts. Even so, using this road is still the easiest way to reach our timber, which was inaccessible to earlier loggers.

"Having to cross Watermelon's land to get to our timber reminds me how Dad bargained for the use of another logger's railroad and got into trouble. After we moved to Wakefield Dad bought an isolated piece of timber, as many loggers were forced to do. He was granted permission to use another operator's spur until his job back in that section was finished. Early one Monday morning Dad pulled into camp, to find that the rails to the landing he was using had been torn up during Sunday. By pulling the tracks, their owner left Dad without transportation for his logs to the main railroad line. His locomotive and Russell cars stood stranded in the timber, useless and unable to haul logs without tracks! In the woods were down-logs ready to be skidded to the landing and loaded. But now Dad had no means of hauling them to the mill. It was too late in the winter and much too costly to build his own railroad into that timber.

"Dad had trusted the wrong person this time. With the end of the winter logging season in sight, our much needed profits lay on the ground in logs cut for winter skidding. Dad was forced to abandon the entire job. For many years, a rusting engine and Russell cars stood deep in the timber, surrounded by rotting logs being slowly consumed by time. Such was the testimony of one logger's faith in another's word. I hope we can keep Watermelon happy or we might have to fly in to this section."

At the edge of the field we came upon a pile of forgotten sleighs aptly marking the spot where our section of timber began. From there a narrow ribbon of black road curved around giant hardwoods. I breathed deeply of the cool sweet forest smell and relaxed until we came to the part of the road which was *plank!*

"What's this?" I gasped. "Another mosquitoes' mire for

breeding bigger and better monsters! I thought you bought a section of high land, not a swamp."

"This is just a soft spot, not wet enough to worry about. These planks spiked to logs laid on the roadbed will give the heavy log trucks a firmer road than gravel would have made through here."

"I'm delighted to know this is just a 'soft spot' and *not* a mosquito incubator."

The planks proved to be wide enough for everyone who came to Tula Camp except me. Riding over those planks for the first time made me perspire, clench my fists, and hope we would make it. Howard gripped the wheel tightly. With an extra squirt of gas, we skidded along the last planks and climbed up the hill to our new camp; Tula! Of course we chose that name. Our camp stood in the adjoining section of the old-time logging and lumbering community of Tula, now only ghost-ridden by the specters of the past.

Water, especially good drinking water, is vital to a set of camp buildings. The men had dug and drilled in several likely places under Howard's direction and had found no water to speak of. To satisfy his own curiosity, and mine, about the willow-twig method of finding water, Howard got a so-called willow-twig expert to locate our well.

"I'm skeptical about this means you've taken to locate the well, Howard. You're sure he's really done this sort of thing before?" I asked.

"Of course. I'm sure he'll find water. He has located dozens of wells around here. I thought of hiring him before we ever began to dig. If I had, we'd have saved a couple of days of pokin' around."

"Well," said I, "I'll believe this witchcraft only when I've seen it."

We left the car and walked to the group of men waiting to dig the well once it was located. They seemed more optimistic about finding a well site today than I was. Only Red scoffed at the idea. "That guy's a fake. Howard's wastin' his

Plank Road to Tula Camp

money and time. If he thinks a well can be found by a willow twig, he's nuts."

"Give the guy a chance and you'll change your tune. I'll bet a dollar. Wanta take a bet?" Baldy asked.

"I've gotta see this ta believe it, I tell ya," was as far as Red would commit himself.

These controversial remarks did not affect the composure of the man who carefully selected and cut a Y-shaped willow branch from a tree near the "soft spot" we had just driven over. As he lumbered up the hill I saw that he was Finnish—clumsy and plodding; a man with a quiet manner whom one would never suspect possessed an unusual talent for finding water with a willow branch.

"Where you want I should start?" he asked Howard.

Howard took him to the spot where the men had dug before unsuccessfully. "Start from here. Any place you find near this spot'll be all right."

The man grasped the two upper ends of the Y branch tightly in his fists. With his fists resting on his knees, he walked around the grounds, slightly squatting and pointing one end of the willow Y out in front of him. In this queer position he systematically covered the area where Howard wanted the well. Howard whispered to me, "If the end of the willow twig pointing out turns down of its own accord while he holds it firmly in that position," he indicated the stooped man, "that is believed to be an indication of an underground flow of water. The stronger the pull, the closer to the surface the water is supposed to be."

The man followed the pull of the willow where it took him. He traced an intricate pattern through the virgin timber, over rotten logs, around trees, up over the knolls, and into the gullies, through the sunshine and shadows of the forest. When the pull became slight and could not be felt, he returned to his focal point and started again in another direction.

After all but giving up, he found a place with considerably more pull. While wondering men and I watched with looks of awe and disbelief on our faces, he moved slowly on. The point of the willow Y turned *down* and *down* with each grotesque step he took. His breathing became a series of grunts—the only sound of which I was conscious. Beads of sweat stood out on his forehead. His knuckles became white and bloodless and showed that his grip never relaxed. Still the willow branch was pulled slowly down against all this man's strength!

Down and down the branch turned. As the man reached the place where the branch pointed directly to the ground, he relaxed his fingers and dropped the branch. It had the bark stripped from the places where it had been gripped so tightly. I gasped! If I had not seen this search for water I should never have believed it could be true.

Red, the scoffer, exclaimed, "If we never find water here, I'd still like to know what forces the branch to point to the ground." He voiced the question I was thinking. Could an unscientific procedure such as this end our search for a well site? Unseen attraction for a willow branch was something for which I have never heard an explanation.

"Why there is a pull at all is a mystery to me. How do you explain this witchery?" I asked Howard in a whisper.

"I know no explanation that is logical. People here simply say, 'To a man who has witchcraft, the pull of the willow tells where water comes closest to the surface!' "

On the spot where the man dropped his willow branch—not over seventy-five feet as a matter of fact, from where the road was to be—the men began immediately to turn up shovelsful of black leaf mold, and, as they dug deeper of ore-red dirt and small stones. Water seeped in rapidly where they dug. Believably or not, as time proved, when Howard installed a pump we had an adequate supply of clear, cold water all the time we logged at Tula Camp!

Other men—saw gangs—began that afternoon to clear the site around the well, while the bulldozer pushed the stumps out of the way and leveled the grounds. While the men worked, Howard and I rested and watched from a safe distance under a giant birch tree with curly bark. We sat on a log upholstered with pullman-green velvet moss.

"That well couldn't be in a better place than where it is, Julie. I think we'll build the bunkhouse on the north side of it and the cookshanty on the south side, both facing the road and the office." Having made his decision, Howard went to tell the men his plans. Late that afternoon, with the major camp building projects underway, Howard and I left the men digging the well and clearing the grounds and returned through Watermelon's gate, passed Tula, and on to our apartment in Wakefield.

On my second visit to Tula Camp, a few days later, a crew of men, not carpenters but handymen who did a great many things well, were busy sawing and nailing the framework of the long low buildings. The locations of the barn, garage and tractor shed, and blacksmith shop, still farther down the road into the timber, were being cleared, too. Arvid, the crippled shacker, who preferred working alone, was busy on the knoll measuring and sawing boards.

I sat on the plank floor of the office and watched the men work. The smell of new lumber, freshly turned earth, and wood smoke from a pile of burning brush was heavy in the air. I looked out upon the camp site, flooded in afternoon sunshine and cluttered with kegs of nails, saws, hammers, sawhorses, rolls of blue-gray tar paper, piles of lumber, and several fat rolls of bright-blue building paper. Propped against the trees angled the windows with their small smoky panes showing signs of previous use. Stacks of iron bunks, bedsprings, barrel stoves, lamps and tin shades, the cookstove, the big black kettle, and boxes of pots and pans, cups and tableware cluttered the grounds. Blacksmith tools, the forge, cant hook irons, chains, horseshoes, bales of hay, harnesses, and all kinds of rigging had been dumped farther away where the barn and blacksmith shop were to be built. Batteries, tires, tool boxes, and oil barrels marked the location of the garage. Near the office an iron bedstead I had never seen before stood against a tree. A heavy chest and a small barrel stove supported springs for the bed. These material things and the blue sky and tall trees composed the real beginnings of pleasant days and cold, silent nights in logging camp for me.

Day after day Arvid patiently sawed boards and nailed them in place. The walls of the office went up, the roof was nailed and covered with tar paper. He fitted the windows in their openings and hung a heavy door on the office. Lastly the handymen shoveled dirt up around the base for protection from cold winds. Howard's voice sounded above the building noises as he gave directions and called orders. Simultaneously while Arvid worked, the other men hammered and sawed, too.

The building of Tula Camp was not done all in one day, but those crisp fall days were so much alike—sunny, warm, happy—that they may as well have been one. Across the world, those September days in 1939 were full of utter despair, anguish, and wretched desolation, but here—a heaven!

Last of all, Arvid put the blue building paper on the walls of our lean-to room built onto the office. Howard and Arvid planned and constructed a counter and shelves in the office and a wooden sink and breadboard for the cookshanty. Then, gathering his carpenter's tools together and putting them in their hand-pegged box, off down the road wandered Arvid, to disappear into the woods. At the edge of our timber by the spring he built himself a shack, where he stayed until he had a thirst at Christmastime.

I decided to stay at camp during the week. Sunday we would spend in town. When I came again to Tula Camp, I brought bedding, window shades, curtains, rag rugs to cover the rough plank floor, and heavy curtains to divide our room from the office. The tingly feeling of fall in the air had turned to a penetrating chill, and the sky was gray and dull. The buildings were all finished; the bunks were against the walls, and tables and benches waited to serve the jacks in the cook-shanty. The cookstove radiated comforting heat, and coals glowed in the blacksmith's forge.

The tractor shed stood with doors open, and horses stamped around the barn. Freshly turned, acrid-smelling sod banked all the buildings. There were no tramped-down paths, but indications where they would be were evident. Howard's log trucks and trailers stood in the newly cleared parking places. The snowplow, with a fresh coat of paint, was a reminder that the lovely fall days were over, that snow would soon be covering the road, and that we would light the fires and keep them fed until spring.

Jacks in Tula Camp

As soon as the cookshanty was finished, Howard engaged his cook, and we made a trip to Marenisco for supplies, groceries, and more blankets. Marenisco is a mill town off the highway south and east of Tula. The town's name was derived from the name of a lumberman's wife, Mary Enid Scott. The first three letters of each of her names make the name Marenisco. The town I found to be made up of homes, many taverns, a few small stores, a boarding house, the Bonifas general store, offices, and the mill. The boarding house, I might say here, is still home for many lumberjacks and millworkers.

The Bonifas store held for sale anything available to sell. We bought groceries, meat, flour, oats, wire, hardware, axes, saws, clothing, and tobacco; and, indeed—if we could have used it at camp—we could have bought an electric clock. All winter either Howard, Baldy, or I was to come to Marenisco for our supplies, and our log trucks dumped hardwood logs in the mill yard behind the store, where they were devoured in a continual stream all day long on the giant conveyor.

Gradually the crew in our newly built bunkhouse became larger and larger. Jacks with full packsacks came up the new path to the office, by twos mostly, to inquire about the prices Howard was paying for cutting. They all eyed me with suspicion until Howard introduced me casually by saying, "This is my wife."

They pulled off their caps. "Howdy, Missus." Thus they

accepted my presence. Howard, each time, talked over cutting prices with the sawyers; if the prices sounded right to the jacks, they would then go along with Howard to the woods to look at the timber. Then, if the timber looked big enough, the sawyers dropped their packsacks under a bunk and came up to the office to get their saw, axes, and files. (Because a crosscut saw has to be used by two men, sawyers paired off into what loggers call "saw gangs.") In a few hours, with their axes razor-sharp and the saw they would use together perfectly filed, the jacks would set out for the timber—Tex going along, barking and sniffing at their heels.

The word about Howard starting winter operations got around; sawyers, swampers, hookers, loaders, teamsters, truck drivers, and gypo gangs (groups of men who cut, skidded, and loaded short pieces and were paid by the number of pieces they handled)—all came to work. Among them were two Swede sawyers, Emil and Ole, who had been sawing together for years and had often worked for Howard's father. There were other saw gangs, too, many of whom had worked for Mr. Anderson, but Emil and Ole were outstanding because Emil was very tall and Ole was quite short. Both were hard workers and friendly.

Ed Taylor, a French Canadian—short and dark-haired, quick-tempered, but affable—came out to work as foreman. To my surprise, he dumped his packsack under the double bunk in the office just around the corner from the doorway to our lean-to bedroom.

I asked Howard, "Does that mean he's going to stay in the office?"

"Sure, the foreman always sleeps in the office of a logging camp. You'll never know he's there. From breakfast till supper he'll be in the woods. After supper until breakfast time he'll be asleep."

Howard was right. When Ed was not working or sleeping, he kept the office swept, filled the stove, and was willing to talk only if I was lonely.

Ed, as foreman, showed the sawyers their strips of timber to cut. He told them the length a log should be cut if it was to be veneer or Number Two hardwood. He put the swampers at their job of cleaning out skidding trails and skidways for the teamsters who hauled the logs out of the woods and who left them sorted conveniently for the loading crews.

Baldy—a check-shirted man, never without his sliver of toothpick in his teeth—who had worked with Howard before and had tricked me into baking pies, drove a truck, as did Joe Rocco, Jake Hoffman, and Melvin Hagstrom. Waino Laakanen, whom the fellows always called "Laggin'," started swamping behind the tractor. During the years he worked for Howard, however, he always did anything Howard asked him to do. He had worked and would work as hooker, loader, truck driver, sawyer, swamper, and handyman.

"Florida" Lee was the cook. "Squint" Keskey worked as cookee, and "Cowboy" Kent was the bull cook. (The nicknames are significant.) Charlie Rabideau came to drive the tractor when Apie went to work for a tractor concern in Marquette. Steve was a teamster Paddy and King would obey. There were other teamsters, but I best remember Steve, the fiery Frenchman. Ray and Clarence, both young fellows, were fast hookers. And then there was Axel who has always worked for Howard—and still does. There were many other men who came and went from Tula camp, mostly sawyers of the old-time variety—homeless men who came to the woods to forget or escape unhappy love affairs, nagging wives, or even prison sentences, I have been told.

The largest building at Tula Camp was the bunkhouse which I toured out of curiosity one afternoon with Howard. Here nearly fifty men slept, played cards, read, and swapped stories of other logging days, or nights spent in the dives on Silver Street. Two big barrel stoves devoured wood by the cord and warmed the men chilled by long days in the open. These stoves dried the jacks' socks and mitts at night and heated the water they used to wash their whiskered faces and their rough, crippled hands.

Cowboy worked steadily most of the mornings hauling chunks of wood into the bunkhouse on a sled the blacksmith made for him. Above the greedy stoves—one at each end of the long building—wires were strung. There rows of socks and mitts quickly dried. Along both walls stood the bunks, single—as decreed by law—but double-decked and covered with dark-gray woolen blankets. The jacks kept their pack-sacks under their bunks, I noticed, or hung them on the wall next to their hats and bright plaid jackets.

At the foot of each bunk stood a bench. Sunlight slanted in through the small windows and made patches of gold on the new plank floor. In the roof were two skylights that were operated by a long pole, and, when opened, let in fresh air—as well as rain and snow! Early in the season there was no need to use the skylight because the bunkhouse smelled fresh and clean; but, as the weeks went by, the smells of fifty men and their drying socks, the odor of wood smoke, kerosene, tobacco, the frail scent of soap, and the heavy stench of manure would need to be replaced by fresh, cold air.

By the large window at the front of the bunkhouse Howard placed a table for card playing, a radio, an unusual assortment of magazines, and a Coleman lamp. Additional kerosene lamps with wide tin shades hung from the long ridge pole. At the end of the bunkhouse, past the identical bunks, I saw the washroom with basins and buckets of water on a long elbow-high table. When the wood racks were full, Cowboy filled these water buckets and the kettles on the stoves. Here, too, at the end of the bunkhouse, stood a cobbler's last and a grindstone.

"After supper, if the jacks have no axes to grind, saws to file, or socks to darn, they expertly resole their boots and make shelves for their toilet articles from prune boxes," Howard told me. "Or, at other times they carve wooden trinkets and spoons." (I have several practical wooden spoons, but I prize most a pint whisky bottle in which one lumberjack ingeniously arranged miniature axes, cant hooks, and a buck saw. Expertly carved and exact in proportion and detail, these tools of lumberjacking encased in their appropriate container represent a colorful occupation, an insatiable thirst, and a pleasant pastime.)

Howard began to turn lumberjacks away because the bunkhouse was full. The sawyers' warning cries of "Tim-ber-r-r" and the sounds of crashing trees silenced the music of the wind in the treetops; the rhythms in the woods were now the sounds made by the horses' harnesses, skidding rigging, the saw blades being pulled through the tree trunks, men's robust voices, and Tex's excited barking—all foreign to this stand of virgin timber but thrilling to me. Tall birch and maple trees, and pungent-smelling hemlock and pine that had stood for over a century in the snow and sun and rain, were cut into logs and piled on the skidways only to be snatched up by the hookers and hoisted to the log trucks. At regular intervals all day long, the trucks pulled through camp on their way to the mills or Brandt's Landing.

The daylight time became suddenly shorter, and the lamps had to be lighted at suppertime. Cold winds blew the brown leaves into piles against the logs and bushes. Empty seed pods rattled with the slightest breeze and frightened away the flocks of birds that gathered to feast upon the scattered seeds. When occasional rainy days kept the men in the bunkhouse, they played cards, read, or began their carving. As the weather became colder the ground froze and snow came in flurries to fall a few inches at a time.

Then one afternoon a storm darkened the cold blue sky. Gently the first flakes fell. Then faster and thicker they came. Along the horizon were broad streaks of pink and rose—the last of the winter sunset. Quickly the pink faded, and the early evening became black and vast. Snowflakes crowded one on top of another, rapidly piling a blanket over the entire camp. After calling Tex in, we dropped the latch on the door, pushed a rag rug against the wide crack at the bottom, filled our stove carefully with solid chunks of hardwood, and put another blanket over our bed. The wind blew. It rattled the windowpanes and the stovepipe. Snow packed all the crevices and sifted down through the open space around the stovepipe. The wind must have whistled and howled all night while we slept, for in the morning snowdrifts were banked high against the buildings.

The sound of footsteps crunching the fresh snow awakened me that morning. Tex barked; Ed let him out, and he joined the barn boss who was on his way to feed the horses. The footsteps became fainter and fainter. The early morning cold was intense. My nose was like ice, but my body was warm under a mountain of blankets. I settled back to what I knew would be only an hour more of sleep, but nevertheless heavenly.

A clear sound rang out over the cold, white-blanketed camp. The cook had hammered the big triangle that hung at the back door of the cookshanty. The sound seemed to be in the air long after the cook had stopped swinging the hammer inside the triangle. The foreman bounced from his bunk—clad only in his underwear, Howard said—and I heard him poke the fire vigorously, break a few pieces of kindling wood, and throw a bigger chunk into the stove. He closed the door of the stove with noisy force as if to say, "Time to get up." Heat soon radiated to our room—as always happened when the curtains were open—and eventually made our corner so hot that I was forced to put my nose close to the blue-papered wall to breathe, or get out of bed.

This cold morning I snuggled blissfully under the warm covers, thinking I was going to have a long sleep. Gradually the heat radiating to my corner began to get uncomfortably warm—then hot—then stifling! I dispiritedly struggled out from the covers. Howard appeared, sly and grinning.

"Not getting up already, are you?" he asked.

"I have to get up or roast." Howard's scheme to get me up had worked without a hitch. Building a hot fire was, more than once, a successful method for Howard to get me out of bed against my will. Most mornings the heat was welcome and cheering. Howard dressed quickly in heavy, warm, dark-gray pants and bright tartan plaid shirt. He jerked his socks down from the line over the stove, where they had been hung to dry. Then he closed the curtains and put on his boots by the warmth of the fire.

My first impulse at the sound of the triangle was always to dive out of the warm covers into the cold and start the race to put on my clothes before the breakfast bell, but today the comfortable bed was hard to leave. The sky was lighter now than when the barn boss and Tex made their early morning trip to the barn, but the sun had not risen as yet. Through the window, curtained with lacy frost of an exquisite fragile pattern, I watched the dark sky lighten. The storm was over; snow transformed everything into unrecognizable shapes, but the result was lovely.

Consulting the thermometer by the office window, Howard whistled and exclaimed, "Colder than Billy Blue this morning. Looks like a good day to log. I hope Charlie drained the tractor last night."

I heard Ed stamping his boots on the floor and whipping the laces through their catches. At this hour between the first bell and the breakfast bell all the men in the bunkhouse were preparing for a cold day in the woods, too. After putting on their favorite type of clothing—chopped-off or rolled-up lumberjack pants of heavy warm wool, bright-colored fancy plaid or checked shirts, several pairs of dry socks, and sturdy, well-greased boots or lumberjack rubbers—they straightened their bunks.

"The legendary lumberjack wore little ta keep himself warm," Ed said. "He worked ta keep from feelin' the cold. In the mornin', not botherin' ta put his shoes on, he walked barefooted ta the creek where he dipped water from a hole chopped in the ice. With this icy cold water he washed up before he ate his salt pork and beans. He wore only his pants and shoes over his long-sleeved underwear while choppin' down the big white pine. The best way ta keep warm, he believed, was ta stay away from the warmth of the stove all winter." There were few jacks at Tula Camp who could verify this legend which Ed told me and expected me to believe. Our jacks wore patched clothes, but they always put on enough so that the icy winds could not penetrate them.

Baldy told me there was always a great deal of kidding and ribbing among the young fellows in the bunkhouse. "In the mornin', after they argue about whose watch is closest to bein' right, they start yappin' about strip counts, scale sheets, and whoppin' big logs."

When the breakfast bell sounded (why they called it the breakfast "bell" I don't know), saying, "Come and get it," in clear ringing sounds, the men rushed out of the bunkhouse without caps or jackets. They poured out so fast that the spring on the door, strong as it was, could not pull the door shut against the hungry fellows on their way to the cookshanty. The path might have been slippery, but their navigation took them to the inviting aroma of hot cakes, fresh sidepork, and coffee, without any such delay as a fall. The last one out of the bunkhouse let the door slam, and he closed the cookshanty door, too. Tex took his customary place on the board platform covered with fresh snow to wait for the food he knew the men would bring to him when they came out.

To satisfy the enormous appetites of lumberjacks who worked out-of-doors all day, our camp tables were set with numerous kinds of food to appeal to all the men and give them ample nourishment for a day of hard work. Each man had his own place at the table. (I ate after the men finished.) Howard sat at the head of one table, and the men had places on either side according to how long they had been working for him. The men moved up to fill a vacant place when a man

They Argue about the Time

left camp, and a new man took a place at the last vacant space. This method kept the men in a group, systematically filling vacant places.

The foreman sat at the head of the second table on the other side of the cookshanty. The tables had many "setups," consisting of such things as containers of sugar, salt, pepper, mustard, syrup, catsup, and crackers. The hot foods were put on just before the men were called to the cookshanty. Six men served themselves from one setup, which made passing dishes of food easier, and all the men were served at nearly the same time. In the summer, when the number of men dropped to thirty or less, the second table was often used as a place on which the cook cut bread or cooled the cakes and cookies.

After the clatter the men made turning their plates and cups right side up, there was only the sound of hungry men eating. The men strictly observed the rule which forbade talking at a camp table except when they found it necessary to ask for food. No talking made eating time shorter, which allowed the cookee, Squint, time to get the dishes washed and back on the tables before it was time to begin peeling potatoes.

When the jacks were not logging, sleeping, eating, drinking, or fighting, they were griping. Their first and last gripe concerned the food. They tolerated poor timber, deep snow, cold, even lice; anything except food which they did not like or had not eaten before. During early logging days, when food had to be toted in to camps, only staples like flour, potatoes, beans, cured meats, tea or coffee, and dried fruits were used in preparing meals. Howard said, when I complained about the frequent tiresome trips to Marenisco, "Totin' in supplies really was a problem then. I did a lot of it for Dad. Before my time, when tote teams hauled enough food for months over long distances, boxes and cartons of food spilled along the trail marked the passage of the tote sleigh. When the horses got winded, the driver lightened the load by leavin' part of the supplies behind, beginning with the supplies the men could get along best without."

As the lumber camps became more accessible, the men were served more elaborate meals. All manner of city foods found their way to the lumberjacks' table. The cooks, however, still served hash as an economical measure; and some hashes have resulted that were epicures' delights. At our camp the cook served fresh fruit to take the place of dried prune, apple, or apricot sauce at breakfast time. The first jacks in to breakfast filled their pockets and left none for those late to the table. So, to prevent that situation and give each his share, the cook made marmalade of the fresh fruit and served that instead.

One morning later in the winter, Emil and Ole came into the office for tobacco. While they were waiting for Ed, I overheard them discussing the meals.

"By yee," said Ole, "that cook's a gude von. Hash, hash, all the time hash."

"Ya," said Emil, "vee have roast beef hash, und roast pork hash, und vegetables hash. Und now ven vee have apples and oranges, the cook makes fruit hash."

Mealtime was no social affair for lumberjacks. Eating was a big part of their business. The cook made sure that all empty dishes were refilled so no jacks would go away hungry. No lumberjacks ever left edible food on their plates. That, like the no-talking rule, is a well-established regulation in logging camps. When a man finished his meal he left the table because there was no reason for him to stay.

Daybreak in Our Timber

The morning after the first snowstorm at Tula Camp I got out of bed to dress while the crew was eating. A winter sunrise of flaming peach and cerise colored the sky and cast a pink light on the smooth drifts. Howard, the first to finish his breakfast, came out into the early morning light. His hair looked red—the kind of light red-blond hair so characteristic of Danish people. A stubble of red whiskers showed on his thin face, and his eyes were bright with anticipation of another good day for logging. Over his bright plaid shirt he wore a dark-gray vest. He preferred laced breeches tucked into hand-sewn boots to cut-off pants and lumberjack rubbers, but he wore either type of clothing, according to the weather.

He lighted a cigarette and came up to the office. Tex followed him a short distance, then returned to wait for the men to come out and go about their work. Baldy and Ed slowly climbed the hill to the office, talking as they came. Charlie went to the bunkhouse for his mitts, cap, and jacket, then went down to the cat-shed. Soon I heard the little motor running; and then the big one drowned out the noise of the starting motor.

Laggin' came out of the cookshanty and pounded his chest as he let out a long howl! No day would be started right for him if he did not let out that yell. It could not be described. I have never heard another sound like it. This strange sound probably emulated the howl I have been told the early lumberjacks let out when they hit the sawdust towns. Laggin's howl certainly expressed pent-up energy, plus lung power.

Laggin' was tall—over six feet. You could tell Laggin' was a lumberjack by his huge shoulders and long arms. Hooking logs, swamping, and sawing may all be arduous, exhausting work, but it builds muscles. Laggin' was Finnish and, like most men of that rugged nationality, full of power. Anyone would have to be built like Laggin' to yell like he did. Laggin' was a good-hearted fellow, but one with whom the men preferred not to disagree, because an argument with Laggin', if carried too far, would end suddenly, in a decisive manner. The name Laggin' suited him very well, because he was much like a stick of lagging—tough, long, and of good quality.

Well, the day after the storm had started right: Laggin' yelling to let out energy, the cat starting without trouble, and now two veneer trucks were coming up the hill to camp. I could hear them. The drivers slithered up beyond the office and stopped at the gas tank. They "filled 'er up" and came to the office to have the number of gallons charged to their accounts. Sometimes they bought a pack or two of cigarettes, a candy bar, or a new pair of mitts. Howard sat behind the counter on his high stool and decided with the foreman which logs to pick up that day. "There's a load of hemlock on the first skidway for you, Joe. The veneer skidway's full—enough for three loads, isn't there, Ed? We'll load you, Baldy, after Joe highballs."

"There's three loads of Number Two piled on the skidway by the second turn out, too, Howard," Ed remembered.

The drivers left the office noisily and started for the woods by inching along behind the snowplow.

Steve, the good-natured teamster, came out of the cook-shanty, followed by the Swede sawyers Emil and Ole. Then Axel appeared with a group of older men who did not eat as fast. When the last man slammed the door, Tex knew somehow that there would be no more scraps for him and he ran after the teamsters. He stayed with them while they harnessed their teams and started down the woods road to the skidding trails. Charlie, on the tractor, passed them in fourth gear and disappeared out of sight around the corner. The sound of the machine came back gradually more faintly as Charlie outdistanced the teamsters.

More truck drivers and Mr. Rogers, the Penokee scaler, came to the office, and steaming men surrounded our barrel stove. If I wanted anything to eat I would have to leave my eavesdropping and go to the cookshanty. Florida made a batch of griddlecakes for himself and me because I liked them best hot off the griddle. I left our office, which overflowed with men. A breakfast fit for a duchess was waiting for me, and, best of all, I did not have to prepare it.

I doubted whether Cowboy, shoveling paths, appreciated the beauty of the drifted snow, but he called a cheery, "Good morning, Missus."

As I entered the cookshanty, Florida flipped our hot cakes onto a platter, and Squint dumped the steaming hot knives and forks out of the flour sack he had dried them in onto the drainboard. Florida finished his breakfast quickly and started making pies for the noon meal. I sat at Howard's place and ate my hot cakes in a leisurely way.

Our cookshanty held two long tables with benches on either side. Between the tables stood a barrel stove of enormous size. A large cookstove, a wooden sink and drainboard, a breadboard, shelves piled high with dishes and supplies, and long hooks on which hung cooking pans and utensils equipped the kitchen at the back of the cookshanty. Behind the kitchen extended a built-on meat house.

In this freezing-cold room, where I went to cut some cheese for a sandwich to take along to the woods, the cook hung sides of beef, pork, and veal, slabs of bacon and fresh side pork, and smoked hams. Boxes of butter and smoked sausages, along with pails of lard, lined the walls. In one corner stood a keg of pickled herring and one of pickled pigs' feet—foods which Howard and most of our lumberjacks considered choice tidbits. Big yellow cheeses stacked on a shelf provided snacks for those who did not like herring or pigs' feet, and cheese sandwiches for me to lunch on in the woods. The butcher's block, a clean white chunk of wood as big as a stump and just the right height for trimming steaks, stood in the middle of the meat house. From a long, low rack by the door to the kitchen an assortment of razor-sharp knives and cleavers hung.

Banked along the back wall of the kitchen extended a high pile of wood which disappeared daily into the cookstove. When Cowboy finished shoveling snow he would replenish the woodpile. Between the boxes of dried fruits, fresh oranges and apples, crackers, and macaroni, and the gallon cans of vegetables, jams, and pickles that banked the side walls by the long tables, the cats played hide-and-seek with mice who grew fat on logging camp diet. Sacks of potatoes, onions, sugar, and flour were piled here, too. Steeped in the smell of the kerosene lamps that hung over the tables and by the kitchen stove and the odors of wood smoke, frying doughnuts, steaming coffee, freshbaked bread, and onions, the cookshanty reeked of that peculiar aroma characteristic of all cookshanties. Activity in the cookshanty continued from early morning until late at night, except for an hour in the middle of the afternoon when Florida took a nap. The silence of the cookshanty then, broken only by the loud ticking of Big Ben, the squeak of a trapped mouse, and the crackling of the wood fire, was so intense that the descent of an icicle to the ground outside the kitchen window sounded like a shelf of china crashing down.

The cook slept in a cubicle built off the front of the cookshanty. Two camp blankets hung in the doorway—closed at mealtimes, but open during the remainder of the day to let in the warmth from the barrel stove. Florida was the only person in camp who rated private sleeping quarters. When the crew was large enough to make a cookee necessary—as it always was at Tula Camp—he slept in the bunkhouse as the bull cook did.

While I was eating, I glimpsed through the window the last of the trucks pulling out to the woods. The loading crews caught rides to save walking. Howard left last, but he took the car, passed the entire crew, and arrived first in the woods. As a rule I hurried to finish my breakfast before he left camp so that I could ride with him to the woods to watch the loading crew and walk along the skidding trails. Today I wanted to walk and enjoy the beauty made by the storm. Yesterday I had lingered over my coffee too long; I walked out later that morning when the sun shone brighter. Catching a ride with a truck going back to the woods for a second load was sometimes my practice.

Across from the cookshanty on the east side of the road, the cat-shed, which sheltered the tractor and served as a repair shop for trucks and the tractor, stood with its doors wide open exposing walls on which the men had hung tools and extra pants. Grease guns and oil cans, I noticed, already cluttered the space under the workbench. The rapid departure of the crew to the timber left no time for straightening or door closing.

Next to the cat-shed stood the blacksmith shop, the special workshop for Paul. This day I stopped just inside his door while he heated and hammered, *clink, clang-g, clongh,* a piece of iron that gradually took the shape of a hook for a jammer pully.

"Good morning, Paul."

"Mornin', Missus."

Paul, who had no time for idle chatter, was the blackest white man I ever knew or have seen since. Only the whites of his eyes and a six-inch length of underwear that showed beyond his cut-off shirt-sleeves gave away his position in the darkness of his shop. The red glow from the hooded forge in the corner and the dim yellow light from two small smoked-up windows gave hardly any shape to the things within the shop. While I watched, Paul laid aside the hook and began shoeing horses. Their restlessness disclosed their position, but Paul with his back turned became part of the shadows.

Paul's shop echoed with the clear ringing sounds of his hammering and smelled of the acid odor of red-hot metal that hissed when he dunked it in a trough of water to cool it. A sickening smell of horses' scorched hoofs merged with the smell of old leather. Parts for harnesses hung in orderly arrangement on racks. A pony harness ornamented with brass studs and ruby-red glass buttons hung beside a worn dog harness. On a work bench under the window Paul had spread out every one of a large assortment of pliers, tongs, and hammers. Now he turned, poked, and found quickly in the pale light the exact tool he needed. It was more than I could figure out.

Paul, fearless of our unusually frisky team, Kaiser and Cap, talked to them gently as he worked. "Whoa there, Cap. Easy, boy. Easy does it." He pared away their hoofs and fitted their shoes. He patted their rumps and eased his big hands down their legs when he lifted their feet into his leather-aproned lap to nail their shoes. "Steady now. Steady, boy."

"Seems to me the horses like to be shod when you do it, Paul." By the way Paul treated Kaiser and Cap, one of our prize teams who were fat and sleek, with silky manes and tails, it was obvious that he liked horses.

"When you get the horses shod, what'll you do, Paul?"

"When I'm through makin' horseshoes, I'll finish some skiddin' tongs, cant hooks, and the jammers. Then I'll make sleds for the bull cook and skiddin' sleighs for the gypo gangs."

"That sounds like enough work to keep you busy all winter from sunrise until sunset, shaping irons at the forge and hewing timbers outside your shop."

"The jammers'll be done today. I just got to bolt them together. When I get the cables run through the pulleys, the bay team, Paddy and King, can pull them out to the skidways. Howard needs this one at a skidway already jammed up with logs, he said."

As I turned to go Paul laid down his hammer. "If you see Howard, tell him I'll have this jammer done by noon. He might want to take it out to the woods this afternoon."

"Sure, I'll tell him," I said, and Paul went back to his forge. (The bearing of messages became one of my jobs from that day on.)

I hiked down the road past the barn, which was the last building in camp and quite a distance from the other buildings. In spite of its being so far from the office, one of the memories I have of camp life is that the steaming pile of manure at the back of the barn indicated a clean barn for another day, and from that same manure pile summer breezes from the south brought to us an aroma—a most unforgettable smell.

Beyond the barn I followed the road that curved through the timber. Howard and I had walked together along this same road when the bulldozer was clearing the stumps and pushing the knolls down into the gullies. The road then was wide, soft, and black and smelled strongly of rotting branches and leaf mold. There is no smell comparable to the acrid smell of freshly turned leaf mold. Later, gravel trucks had covered the road with layers of stones, and log trucks, leaving the woods with their heavy log loads, soon packed the gravel into a hard surface.

This day the road to the woods—a fresh white trail showing truck and tractor marks and men's and horses' footprints—curved through a forest alive with logging activity. It led past tall birch whose curly bark held fresh-drifted snow. Along the roadside I saw trees whose boughs bent with the weight of heavy snow. Mounds of snow adorned the fungus shelves and the chunks of snow slid from high-up hemlock boughs and fell, "plop," to the snowdrifts. Tex came bounding down the road to meet me. We walked on together to where the loading crew was preparing to load a truck with veneer logs.

Strong Day's Work

On our walk to the woods Tex and I retreated knee-deep to the drifts along the road to let log trucks wheel by us. These loads were evidence that the loading crew had begun their strong day's work when the sun was still hidden beyond the horizon. As we came upon the first skidway a mammoth ball of flame rapidly rose above the treetops. Radiant sunshine flooded the woods, and shadows became sharp and dark on the pink-tinted snow. The temperature seemed to rise with the appearance of the sun. Tex barked joyously at the sight of the hookers. They had no time to play with him, for at that moment a truck ground to a stop under the jammer. The driver Domenic, impatient to get started on his haul, grabbed his cant hook and leaped nimbly to the top of his cab ready to load.

"Let 'em come, men," he yelled.

The hookers, Ray and Clarence, dropped the trailer stakes on the loading side and put two skids at an angle against the bunks to prevent the logs from banging against the tires and the framework.

"Pile on the logs," called Ray.

"Here they come," yelled Clarence, indicating that he, too, was all set for hoisting logs. And the logs went on—faster than can be told!

Ray and Clarence pulled their pups (the hooks that grab the log) back to the log Domenic selected with a motion.

With the sucker line (a rope—sometimes a chain—with which the hookers guide the logs to their position on the load) in one hand, they placed the pups on the ends of the log with the other hand, still retaining hold of the sucker line. When the pups were in place on the ends of the log, Clarence signaled the crosshaul teamster with a wave of his hand. Steve's flick of the reins started his team away from the jammer, and the log rolled toward the truck.

Fastened to the evener behind the team was the cable that ran through a pully at the top of the jammer boom and tightened the pups in the logs. The team strained forward, pulling the log up the skids. The hookers steadied the log with the sucker lines while it hung above the bunks. At a signal from Domenic, Steve slowly backed Paddy and King, and the log dropped into its place. It fell at an angle with a thud that jarred chunks of snow and ice loose from under the trailer. Domenic jumped from the top of the cab and deftly rolled the log into position with his cant hook. I marveled at the easy motions he used to secure the log against the stakes.

With the first jammer log in place, the hookers jerked their sucker lines hard to loosen the pups, while the team jogged back at a fast pace to the jammer and turned around. The cable, freed from the evener by Steve, was pulled back by the hookers when they began the loading procedure again by placing the pups in another log. I watched them hoist up another log and ease it down beside the first log. Then they loaded a face log and dropped a wedge log in place beside it to secure the bottom layer. The crew piled on another layer of logs in the same way; then they bolted the stakes upright against the logs.

Clarence hooked the cross-chains to the stakes to hold the bottom layers on the trailer. Now the men began hoisting on the top of the load. I watched heavy, big birch logs swing high in the air to ride up and over the stakes. Domenic, who selected the logs which he thought would fit best as face logs, wedge logs, and jammer logs to balance his load, swung onto the top logs as they came over the stakes and rode them to their niches. As he pried the last smooth maple tightly into its place, I heard him yell, "That's all."

The hookers kicked the skids out of the way to use them again and clamped the load binders. (The binders fasten the wrappers of chain which hold the entire load of maple and birch veneer to the trailer.) Domenic tossed his cant hook to the ground, swung down from the cab with the agility of an acrobat, tested the load binders and chains, then climbed into the cab to start the motor.

"Be seein' you guys," he called, as he carefully pulled out from under the jammer, slowly gained speed, and disappeared from sight around a curve. The growl of the truck leaving the woods came back to us whenever he reached the top of a hill; it died away as he coasted into the gullies. Finally, we could hear the truck no more.

Clarence puffed on a cigarette and Ray played with Tex while the next truck pulled under the jammer. A whiff of tobacco smoke came to me through the fresh, cold air. Joe Rocco, a smiling, happy Italian, swung out of his cab and yelled to the crew, "Pile 'em on, boys. I'm gonna break the record today." Clarence flicked his cigarette into the snow and grabbed a sucker line and pup. In a matter of minutes another load of veneer was on its way to the mill at Mellen.

I exclaimed about the size of Joe's load as I watched it sway and careen down the road. "Gee, that's a big load! About the biggest I've seen leave the woods this winter."

Steve, bringing a log to the nearly empty skidway, heard me and said, "That's just a small load compared with the loads I've seen some loaders pile on sleds. 'Sky-high' loads they were. When loggers used horses for hauling and iced the roads, they piled the logs so high all ya could see when ya looked up at the top was sky all 'round."

"I've heard about sky-high loads, Steve. Tell me, how many

"Sky High" Load

teams of horses did the drivers use to haul such big loads as those?"

"Well, Missus, it sometimes took three teams to get 'em started away from the skidway; but most of the sleighhaul roads out of the timber were downhill all the way to the landin's—and crooked, too. One team—like Paddy and King here—could pull an awful big load, once it got started. Road monkeys iced them roads and cut deep ruts around the curves to guide the runners. Sometimes loads tipped over on the curves. Teamsters got killed and bashed up once in a while. Horses got shook up some, too—and killed. 'Twas dangerous business—taking a sky-high load to the landin'. Couldn't get me to do it today for no money, but I hauled plenty of 'em in my day. I was no more than a kid then. You're right, Missus, Joe's load was big, but you shoulda seen sky-high loads. They was half again as big."

Steve touched his team with the reins, and they jogged down a narrow skidding trail to drag out more veneer logs to fill the depleted skidway. Clarence and Ray pulled the jammer boom back, fastened the guy lines to sturdy saplings, and hiked down the road to load hemlock from another skidway. With Tex at my heels sniffing footprints, I followed the hookers to find Howard and give him messages from the cook and Paul. I found him by following the sound of his voice, which I heard above the other logging noises when he yelled to a truck driver, "Take it easy. Your load is heavy on the jammer side."

All the logs from Tula Camp were not hauled out of the woods without accidents or delays. One morning when I had stayed behind in the cozy office to complete a sketch by the warmth of the office barrel stove, I heard the car stop with a squeal of brakes. I wondered why Howard was in such a hurry and looked out to see him helping Domenic slowly climb up the path to the office. Domenic held a wad of spotted white to his head. His face, ghastly gray and streaked with crimson, told me that the accident everyone had hoped could be avoided had suddenly happened. I drew in a deep breath and braced myself for what I sensed would be a gruesome sight and a bloody job. As they crowded through the doorway together, I saw that Howard's face was as colorless as Domenic's. And when he spoke his usually loud and clear voice was husky, low, and strained.

Hoarsely he said, "Get out the first-aid box on the counter."

Domenic dropped his arm from around Howard's shoulder and slumped onto a stool. As Domenic dropped his hand from his head, his blood-soaked handkerchief rolled to the floor. I handed Howard the first-aid box and took one look at the purple gash at Domenic's hairline from which oozed crimson blood. The faint but sickening smell of blood reached me. My stomach turned a flip-flop, and I leaned weakly against the counter. The next second I slid, unconscious, to the floor.

"You went out like a light," Howard told me later, when he came back to see how I had stood the shock and to console me for being such a weak-kneed nurse's aide. "I thought Domenic was going to pass out, too, but after I got you straightened out on the floor I cleaned him up with cold water. That revived him, and he felt pretty good on the way to the hospital."

"You didn't look very healthy when I last saw you. What revived you?" I asked.

"The cold air, I guess, and the fast ride to the doctor."

"Where's Domenic?"

"Oh, the doctor took a couple of stitches, and I brought him back. He just went through camp with his load of veneer. Nothing wrong with him now."

"What happened, Howard?"

"Unavoidable thing. Ray slipped and let go of the sucker line. The loose end of the log clipped Domenic as it swung around. The bad thing is that this accident won't be the last for a while. We can expect a couple more; then our bad luck will be over."

"What do you mean 'we expect a couple more' accidents? Why?"

"Logging-camp accidents come in groups of threes. Watch and see."

Unusual as it may seem, things did happen at Tula Camp in groups of threes! Domenic's accident was followed by two more. The men did not think of themselves as superstitious, but they doubled their watchfulness after Domenic's surprise accident. Even that precaution of added carefulness did not prevent a teamster from spraining his ankle and Ole from cutting his hand on his keen-edged ax a couple of days later. If the men had not been half-expecting these accidents, it is probable that they would have been much worse. When the last of three accidents occurred, all of us found the tension relieved.

The day Ole left the office with his hand bandaged we were discussing the superstition about accidents happening in threes. "I didn't know jacks were superstitious, Howard."

"They aren't really. Just a certain few things are ill omens to them. Men have been tempted to 'lay in' on the day they broke their bootlace, because to them that is the sign of a pending accident, probably involving them. After an accident some jack in the bunkhouse can always recall having just broken a shoelace. If a lumberjack finds a knot in the rigging chains, he'll yell with gusto, 'Who the hell is next? I be damn sure it ain't me.' To him the knot is a sure sign that someone is 'goin' down the road' and he doesn't want to be fired. When he yells *his* intentions of staying on, that breaks the spell of the knot as far as he is concerned."

I noticed that most of the men seemed to realize that logging is an extremely dangerous occupation and were always watchful for conditions that might cause accidents. Most of the accidents at Tula Camp were slight and resulted only in short delays and personal inconvenience. Compensation and accident insurance relieved the financial aspects of these accidents.

But concerning compensation, there is Curly Angus' story.

He worked hard for Howard sawing logs. One day he crushed the end of one finger between two logs. Ed took Curly to a doctor, who found the tip of the finger so badly mutilated that he amputated it at the first joint. Curley Angus lost only a few days on account of his unfortunate accident. After some time he was allowed compensation of fourteen dollars a week for twenty weeks for the loss of his finger tip. Each Friday, Curly received a weekly check which he cashed at the nearest tavern; and, at the same time, he had a drink or two, with the result that he was not always in condition to work either Saturday or Monday of each week for those twenty weeks. I asked him, one Tuesday morning, "Where have you been, Curly? You didn't work yesterday."

I received this prompt reply. "Vell, you see, Missus, my finger vas amputationed. Now compensation set in, und I can't vork so much."

Insurance companies sent representatives to Tula Camp to give talks and show movies in the bunkhouse for the purpose of demonstrating to the lumberjacks ways of preventing accidents and of keeping them alert for potential accidents. In spite of all the precautions employed by the insurance companies, the jacks, and Howard to avoid accidents, some men did land in the hospital for one reason or another. Both the truck drivers and the jacks working in the woods suffered their share of misfortune. Sawyers cut off fingers, slashed their boots, and cut their legs and feet. Branches broken from falling timber—the "widow-makers"—gashed heads and bruised arms and shoulders. Logs rolling unexpectedly from decks or loaded trailers broke arms and legs. More than one driver turned a corner too sharply, tipping the trailer and logs over and breaking the stakes and snapping the binders. Occasionally the cab tipped over, too, denting the fenders and doors, shattering glass, and jolting the driver.

The only delay these accidents caused was the time spent in reloading the logs and repairing the cab and trailer. Howard's woods roads were all one-way traffic, which made a complete circle in the timber in order that trucks returning empty could wheel and not have to slow down or turn out for loaded trucks leaving the woods. By careful driving, loads could be hauled out of the woods on what normally would be the return road when an overturned load blocked one part of the road. Howard then sent the tractor, a jammer, and the loading crew to right the truck and trailer. If the damage was not too great, the crew reloaded the logs, and the driver started on his haul again to the mill or the landing.

The cautious drivers took their loads slowly out of the woods, up past the office, down the hill over the plank road, through the field and Watermelon's gate, and onto the highway. Some drivers took their loads of Number Two hardwood to the Bonifas Lumber Mill in Marenisco. If I happened to be picking up groceries for the cook at Bonifas Store when they pulled into the mill yard, I watched the drivers unhook the binders, slip off the wrappers, and trip the stakes with their cant hooks to let their logs roll onto skids to the hot pond or onto decks in the yard.

Watching the logs roll and pound down from the load with a deafening noise, I realized how dangerous this part of hauling logs is. A log loosened from a load without warning could kill a man in an instant. I could see why log haulers learned to be alert and quick, because unpredictable things happened when the stakes were tripped. A burly knot held a birch log, balanced and swinging precariously, high on a load Gerke dumped one morning at the mill. With quick jabs from his cant hook and agile retreats to safety, Gerke brought this potential killer pounding crazily down with terrific force to lie inertly awaiting the last record of its long life as measured by the scaler's rule. With their scale sheets tucked above the rear-view mirrors and loose chains jangling on their bouncing and swaying empty trailers, the drivers wheeled back to camp to flaunt their scale sheets in the faces of other drivers. Good-natured rivalry over who could haul the biggest loads was continuous at Tula Camp. This rivalry provided a constant source for kidding and conversation.

"Head Push"

Once the first log trucks left the woods from our Tula Camp in the morning, others followed in rapid succession. It was an epic battle that Howard—"head push," the crew called him—conducted as boss of his logging army. All day long the sawyers felled the trees and sawed them into logs. The swampers cleared the skidding trails. Teamsters, when they were not busy on the crosshaul, sorted and pulled the logs to the skidways. Hookers decked the logs for fast loading. And when the trucks came wheeling back empty, the loading crew quickly had them loaded again. Some drivers made their day count mightily by completing three trips to Marenisco, or five short hauls to Brandt's Landing. Penokee Veneer truckers generally made two trips to Mellen in a day.

Mr. Rogers, Penokee's scaler, scaled the veneer loads at the skidways in the woods and gave the sheets directly to Howard. To satisfy the truck drivers, he made extra copies for them to compare.

"A good egg," the log haulers called him. I remember this short, stocky Scotchman with deep-set, bright-blue eyes and ruddy complexion very well. Mr. Rogers always wore a hand-knitted muffler tied close under his chin to cover the tips of his ears, and pulled the earlappers of his Scotch plaid cap down to meet the roll of the muffler. Otherwise, he dressed much like the other men in a warm jacket, heavy pants with rolled-up cuffs, and sturdy footwear.

I thought the distinguishing thing about this likeable man was that, although he was the father-in-law of President McIvar of Penokee Veneer Mill (one of the largest veneer mills in the United States; it produced much of the veneer used in vital war products) and although he could probably have had a soft job inside that mill, he still chose to work in the woods. As he said, " 'Tis invigorating, lads, to be oot of doors all day in the woods. 'Tis that, me lads."

It was he who made me aware of our comfortable life at camp. "A bien abode, 'tis what you've got here," he said whenever he warmed himself at our barrel stove.

The veneer company Mr. Rogers worked for bought our choice hardwood logs and hauled them to Mellen at their own expense. All other logs and forest products we hauled to mills and landings with our own or hired trucks, and were paid for them accordingly by the paper companies, lumber mills, and the railway companies. Because the hemlock haul to Brandt's Landing on the outskirts of Wakefield was the shortest, the drivers hauling hemlock piled on tremendous loads. Scale sheets were kept of all the logs decked there, but the final count was taken in the spring when the bark was peeled to be shipped to leather tanneries. There the tannic acid was extracted and the peeled logs were shipped to mills in Wisconsin to become paper pulp.

Besides felling and hauling log timber, Howard's valiant crew also produced other forest products to supply the mines and railroads with the timbers they needed. Lagging, ties and mining timber were all short stuff cut from smaller trees by gypo gangs, individual piece-workers, and shackers. It was a part of Howard's careful planning in his struggle to win success, not only financially but also in his career as a good logger. Along the woods roads I watched the gypo gangs hoist this small stuff onto piles which grew higher and longer as the winter progressed. When the proper time came, these products were hauled—the ties to a railroad landing, and the lagging and mining timber to the iron ore mines in the Upper Peninsula for use underground.

Some piles of short stuff bordered winter roads where the piles grew, diminished, and were replenished continuously during the hauling season of the cold winter months. Other piles along roads that were planned for summer hauling purposes became heavily covered with snow; it formed overhanging roofs which made the piles look like small Swiss chalets whose grotesque shadows picturesquely adorned the smooth, untrampled expanse of snow-covered roadway.

Loading lagging, ties, and mining timber—usually done without the aid of a jammer—involved laborious work. Cutting, loading, and hauling these small products by the piece meant that the wages the gypos could earn were limited only by their endurance. The men who wanted to work long hours did it; and, by so doing, they earned more money. That right to earn according to one's ambitions and ability! It is one of the freedoms of this country. These vast woods are the scenes of real exercise of our American genius.

Howard employed a great number of young fellows who bargained separately for their jobs, lived up to their promises, and were agreeably inclined to do extra work, which always had to be done at unusual hours at Tula Camp. Whether the job Howard asked them to do was their special work did not matter to them. If they did many different jobs, that was no concern of a union, because they were not "packing a card" as many jacks had begun to do. From the conversations I overheard in the office, I could tell that the young fellows felt considerable pride in knowing they were working on a logging job that was making money for the boss. No one complained of long hours of strenuous work. Although Howard worked just as hard as he had the winter before we were married, his face showed little fatigue, and he laughed frequently. Obviously he was enjoying the strong day's work. As weeks passed, our profits increased, and Howard gave his men credit for making things click.

Respect for Howard showed in everything that was accomplished. To Howard the lumberjacks gave the credit for the timber being a "good chance." And rightfully so. It was

Howard who had looked (a word loggers use instead of "cruise") the timber, literally walking the soles off his boots searching for a piece of timber with trees that would pay. And here it was, a good stand of veneer, sound Number Two hardwood, and plenty of smaller trees for mining timbers, ties, and lagging!

I could see by Howard's restlessness to be in the timber with his men that logging, being a way of life as well as a business, was irresistible for Howard. Here, at Tula Camp, was a challenge to his hard-earned knowledge of logging methods, his way of handling men, and his ability to deal with the problems this particular stand of timber presented in swamps and high lands.

Howard has a cruiser's eye. This ability to calculate the kind and value of the logs a tree will make while it is still green with life is as valuable to him as his ability to tramp around in strange, unblazed sections and come back to the place where he entered the timber.

Being able to sense directions and use a compass are just small accomplishments. But I would never go into a section of Michigan timber without following Howard or a familiar logging road. Howard learned what he knows about timber and logging from his dad and by bitter experiences of his own. He can bargain for stumpage and not get stung. He knows the requisites of a set of logging-camp buildings and how to lay them out to the advantage of all his crew. He enjoys the stimulating environment of pungent-smelling woods and crisp cold winter days in tall timber. He is, out of necessity, a far-sighted man whose best asset is his youth. He believes in advancement in methods of logging and has always been able to rig up a new, practical, easily managed device to replace old equipment and do the same job more cheaply and quickly.

The men at Tula Camp knew these things about Howard, as they revealed by their friendly comments while I played blackjack with them in the office during the long winter evenings. I heard the same things as I lunched with them afternoons in the cookshanty, or chatted with them on the skidways.

Ed said, "We respect him also because he's fair with us and doesn't expect anyone to do what he can't or won't do. He can fell a tree and saw logs as well as any of his sawyers. He can scale them as accurately as Mr. Rogers can—or any scaler, for that matter."

Steve said more than once, "Howard can drive any of his horses. Some other boss would be scared of a wild team like Cap and Kaiser, but not Howard." Charlie told how Howard had driven the tractor in places from which another driver had turned away.

I never saw a loader who could load as fast or balance a load as well as Howard. The drivers agreed with me that he was a first-rate top loader. No load of Howard's logs ever stayed in the woods for want of a driver, either. Howard could wheel out of the woods with a trailer of logs just as expertly as any driver he ever hired.

When no one could find the trouble that had caused the tractor or trucks to stop, Howard could.

Besides, he was generous with Christmas bonuses. On several occasions I saw him give away his own shirts and socks and mitts to jacks when the need arose. Once I overheard him ask of a thinly clad lumberjack for whom he had no work, "Could you use a pair of mitts, Jack?"

"What I got'll do."

"Here, these got too small. Left 'em in the rain. I've got no use for them." I heard Howard slap the mitts together. The lumberjack mumbled what sounded like a thanks. "You're sorta like your old man Hungry Louie. Ya kinda remind me of him. Used to work for him at Planter. I'll be gettin' on now."

And so the jacks and the loading crew were generous toward Howard in their own ways. When snowstorms closed the roads during the day, they would rig up a headlight for the tractor and would plow through the drifts at night, working to have the roads open by morning so the men could go to the woods as usual. Many nights after supper the fellows who could drive the tractor gathered in the office to decide what hours they would take their turn plowing snow. The one who went out first, climbed into a big sheepskin-lined coat of Howard's—the only thing that could keep a fellow warm while he rode the slow-moving, completely air-conditioned tractor snowplow.

Because the woods roads were narrow, they were difficult to plow for lack of room to pile the snow. The drifts across the open field between Watermelon's place and camp were deeper and often higher than the plow which made more plowing necessary there. But on one typical occasion, out of the prevailing good will and devotion to the boss, one man plowed the snow from the woods roads, then another climbed into Howard's sheepskin and roared away on the tractor to open the road to the highway.

Playing cribbage or poker in the wan, flickering light of our kerosene lamp and soaking up the warmth of our barrel stove fire, Howard and I waited in the office for the return of the man plowing. Howard anxiously kept track of the hours the man was gone. Looking at his watch and peering out through the night filled with whirling snowflakes, he would repeatedly say, "Charlie ought to be back by now. Been gone two hours. Do you suppose he's broke down in a snowdrift?"

To reassure Howard, but not knowing for certain, I would answer, "I think I hear the tractor now." Together we would stand in the open doorway. The cold night air would ruffle our hair, snowflakes would wet our faces, and the heat of the stove would warm our backs while we watched and listened.

At last the drone of the motor sounding far away, but becoming louder and louder—at last, too, the sight of the light from the tractor as it appeared around the curve beyond the barn, relieved our worry. The tractor's noise and its light meant that all had gone well.

At other less lucky times the feeble wavering beam of light from Charlie's flashlight flickering through the silent snow-filled darkness meant a breakdown or that he had buried the tractor in a drift too deep to move. If Charlie came for help, Howard would pull on his boots; and together they would bring the tractor back to camp. Then they would come back to the office heavily covered with snow and shake it off onto the stove. What a sizzle-up of steam—the drops of melted snow hissing and dancing around on the glowing metal until the drops went off into nothing!

While Howard and Charlie talked of finishing the job in the morning or fixing the tractor, as the case happened to be, I would make a pot of strong, steaming coffee from water continually boiling in my copper kettle on the stove. The hot coffee was warming and tasted good to me as well as to the men chilled from hours of exposure to snow and winds and cold.

Snowplowing kept us up far into the night. When Charlie and the others made up their hours of lost sleep, I do not know—Howard never did. Each morning he rolled out with the men, ate with them, and went out to the woods at the same time they did. Perhaps that is why the men never grumbled about extra hours and work. Howard was willing and able to be one of them in the big battle. This arduous and long day's work was a vital part of those wonderful years at Tula Camp.

Our Big Pine

The short winter days came fast, one upon the other. How really pleasant was the life at Tula Camp did not enter into my thoughts until retrospection eliminated a few minor discomforts and emphasized the major beauties and pleasures of the natural life of the woods.

One day I remember well began like many other winter days. The buildings and trees in the bright sunshine cast deep shadows on the sparkling snow. The warm sun had melted the ice on the paths, and water dripped faster than usual from the icicles on the sunny sides of the buildings. Perched on the high stool behind the counter in the office, I watched a flock of rosy, crested birds pick up oats spilled along the road.

As Howard came up the path they fluttered to a maple sapling near the window. He entered the office, bringing with him a refreshing draft of cold air. "Want to go with me while I count logs on a strip a saw gang just finished?" Howard busily looked for a colored marking crayon in the desk.

"I certainly do! Just give me time to get my mitts and put on my jacket." In my haste I tipped over the stool. The clatter frightened the birds, and they flew back to the scattered oats.

Whenever Howard asked me to hike with him over the woods roads or through the timber, summer or winter, I went if I possibly could. This day I lost no time finding my jacket.

Tex, who had followed Howard in from the woods, nuzzled

first my hand and then Howard's, not knowing, I suppose, just whom he should follow as we walked to the car.

To save time, we drove to the strip.

Tex rode in his usual place on the fender with his front paws hugging the headlight.

Howard parked the car in an empty skidway, and we began our hike through the freshly cut timber following the sawyers' narrow trails.

To record the number of logs they have cut, the sawyers slash a chip from the side of each stump; then they mark on the stump with crayon the number and kind of logs into which the tree has been sawed. When the foreman checks the count, he makes a record of the sawyers' logs and also marks each stump when he has located the logs. Checking in this manner meant that no logs are counted twice. Counting the sawyers' logs is not easy except on mild days, as the strip counts I later recorded could testify.

When the weather was stormy, the foreman's paper grew wet with the snow that blurred his figures. The cold numbed and stiffened his fingers. This handicap caused his numbers to be awkward and hard for me to read. But strip counts made on nice days were crisp, legible records of many strong days' work.

The particular day I went, as I say, with Howard he finished counting the logs in a short time. As I followed him from one stump to another, I saw in several places the charred remains of small fires the sawyers had built to warm their stiff fingers on cold days. The smell of wood smoke hung heavy in these secluded spots surrounded by pungent-smelling hemlock slashings and cedar boughs.

I thought of the men who had warmed themselves at the flames of these once-cheery open fires. Hale and hearty men they were, with muscles hard as maple saplings, legs sturdy as oak, and hands gnarled as burly birch. Well able they were to endure the physical discomforts and the many hazards of lumberjacking, but still unsmiling, silent, and grim, they kept their ways. How appropriate, I thought, that they should choose to spend their lives in vast forests just as silent, just as gnarled and strong!

Because it was still early that afternoon as we emerged from the timber, Howard suggested, "Let's go on to next year's loggings. It's quite a distance, but I'll make a trail you can follow easily."

I agreed to go on.

I staggered after him as I tried to walk in his wideapart footprints—his idea of an easy-to-follow trail.

Tex took side trips of his own, dashing under cedar boughs and through the slashings to chase chattering squirrels and scatter flocks of chickadees, snow buntings, and cedar waxwings.

We came at last to what Howard said was the edge of our timber. Indicating a line through the trees with a downward wave of his arm, he said, "This is where the crossroad will connect two parallel roads from camp. There's only one deep ravine we'll have to fill."

We tramped around, exclaiming about the big birch and the tall, straight maple. Then I saw a huge white pine whose trunk was so big that Howard and Tex became completely hidden behind it.

"Look, Howard!" I called excitedly. "Do you see the big pine I do? I can't believe my eyes! Are you sure this is the right section?"

Howard appeared from behind the pine and exclaimed, "It sure is a big pine!"

He circled it again. He stopped to get his bearings.

"Sure, it's on our section. Where we're standing now is a good five hundred feet from the line."

To see the top of the pine, we strained our heads backwards and shaded our eyes. Its height and girth amazed me.

In comparison, the hardwood trees around it looked like stunted underbrush.

This white pine had its face in the heavens.

Behind the barn at Tula Camp, only a few days before, I had found two old stumps—mammoth, moss-covered black stumps. It was clear enough that many years ago big white pine had been stolen from this region. To say stolen is, in a certain sense, incorrect. In the days when white-pine logging was at its height, timber did not need to be stolen. It was simply—well—there to take in plenty.

"It was far better in those days," Howard explained to me when I asked him about the stumps I had found, "to log a 'round forty,' which was what many timbermen did. A lumberman would buy a forty, the smallest tract of land that could be bought in those times; then he would order his foreman to log 'around forty acres.' This the foreman did by loggin' the north forty adjacent to the legally owned forty, the one adjacent at the east, the one to the south, and the one to the west. Hence, logging a 'round forty' was cutting trees never bought!

"Sometimes only a few select pine were cut on the surrounding forties. If these logging operations were discovered, the only legal proceeding involved was the payment of the estimated value of the forties involved, usually never more than one dollar and twenty-five cents an acre. Even a few pine trees like the one here were worth the fine! Logging a 'round forty' seemed quite a joke to many lumbermen," Howard said,"—especially in the days when white pine was plentiful. Today, with no large tracts of pine left in Michigan, that anybody could get by with such a procedure is no joke."

I thought all the big pines had been cut; and I had never expected to see a giant like this one that stood before us in our aged tract of timber.

Howard's face betrayed his pleasure at finding such a beautiful big tree.

That he had missed it when he had cruised the section was certain, because a pine such as this would not be passed by calmly or easily forgotten.

As Howard appraised it now for its beauty, he mumbled

Notching a Burly Birch

a jargon of figures which I did not understand, and he calculated the number of logs its great height would cut, and how much each log would probably scale.

I did know that a high price could be obtained for pine, smooth white pine that was the result of many centuries of growing in the quietness of an unmolested forest. The magnificence! The struggle, for long long years, to push through the thick carpet of decayed leaves—to resist the weight of heavy snows—to find the radiant sunlight above the maple and birch saplings—to reach up and up; then, each year putting forth new, pale-green needles and dropping fragrant cones, sticky with pitch, from tiptop branches, to become this majestic tree—an enthralling sight!

As we left to return to the car, I realized that perhaps I would not see this giant pine again until it passed by the office, cut into logs and loaded on trucks.

I turned to look again. The crown of this pine etched a dark jagged pattern on drifting clouds in the pale-gray sky. The branches farther down the solid trunk became a deep purple-green mass broken only by a missing limb, probably torn down by a falling hardwood tree which the wind had uprooted when the pine was small, or ripped away by a bolt of lightning in recent years. The lowest branches spread out, drooping slightly as if protecting the hardwoods beneath. Its sturdy trunk, rough of bark and piney-smelling—a hundred feet of straight, limbless fiber, and rooted solidly in rocky, snow-covered ground—was an undisputable manifestation of boundless strength and inestimable age. A mingling of minor tones in music came to me from the lofty cone-laden branches as the wind blew. The wind ceased. The song died away. Even so, this great pine would leave its memory.

Interrupting my thoughts of this centuries-old pine, Tex nosed my hand, whined softly, and reminded me that Howard in his proper preoccupation had gone back along the trail we had made. To catch up to him I would have to hurry.

Inspired by this first hike, I came back to the big pine in

spring, in summer, and again in the fall and winter. The hardwoods put forth crimson leaf buds in the spring. Gentle falling rains and warm winds blowing beneath the pine laid bare a moss-covered log and a slate-gray rock that cut through a bed of slippery needles and brown cones with wide-spread scales. In the fall the green foliage of the hardwoods flamed into brilliant color. Winter winds scattered the fallen leaves and drifted fresh snow over rock and log, leaves and cones. Yet the big pine never visibly changed. Its song continued on the winds.

Part Three

Skidding Trails

Part of the time while Howard and the men worked in the woods, I worked, too. With the installation of the gas pump came the necessity of keeping a record of where the gas was used. Because I spent more time in the office than anyone else, Howard suggested, "Julie, you could easily keep track of the gas. Just mark down in this book the number of gallons used by each driver." I was in a pleasant mood and agreed.

The truck drivers began asking me for van whenever they got gas, and then the jacks started buying their cigarettes and candy from me, too. Gradually I was forced to keep the van records. One evening at the end of October, when Howard was figuring our logging expenses for that month, he said in his most tactful voice, "It's just a simple matter to total these gas and van accounts, and you would save time for me. Will you do that for me tomorrow?"

"Sure I will." That was the first of many afternoons I spent totaling columns and columns of figures. One promise led to another until at last I had agreed to do all the bookkeeping except figuring the lumberjacks' time and pay. Howard did that in less than several hours.

I learned to make out insurance, unemployment, and social-security reports. I totaled scale sheets, I wrote checks to pay bills and the men; I sent statements and kept the bank account balanced—on paper. This bookkeeping would have been easy for an experienced person, but for me, whose ability is along other not such exacting, confining lines, it was the hardest work imaginable. Frankly, I hated it, but I did my best; and Howard appreciated my effort because what I did allowed him to be in the woods where he was needed most and liked best to be.

I found most pleasure spending the afternoons in the woods with Tex for company if I could coax him away from the sawyers and Howard. On one occasion, when I took our newly acquired movie camera with me, I was lucky enough to get some excellent pictures of the rosy, crested, purple finches that flocked in great numbers along the woods roads. These birds are unusual and seemed to have been misnamed. When I asked a friend of ours, a well-informed ornithologist, about these friendly, pretty birds whose rosy-feathered brightness showed so vividly against the snow, he told me that he thought they were purple finches.

"*Purple* finches!" I exclaimed. "But these little birds are *rosy*-colored. They can't be purple finches."

He laughed and explained, "The name 'purple finch' really doesn't suit them at all. The male purple finch is for two seasons nearly as drab as its mate, a decidedly sparrow-like bird. During the third season the feathers on the head, breast, and rump of the male finch become a bright raspberry or rose-red—not purple as the name suggests—and the feathers remain that color thereafter."

At first I thought these birds might be pine grosbeaks, but they were not so large as grosbeaks. The small size of my birds, their slightly crested heads, their thick bills, the bristles at their nostrils, and their forked tails identified them as purple finches. Beneath the wintry sky they swung on dried stalks of goldenrod above the crust of the snowdrifts. They ate the seeds of mountain ash, of bittersweet berries, and of weeds, and picked up oats spilled by the barn and along the camp roads. One flock of finches found the branches of the maple sapling just outside our office window their favorite perch. Their cheery call, always the same sweet notes, gave me company when I stayed alone in the office to struggle with my logger's book or to work on a sketch of baffling skidding rigging. In the woods they fluttered from weed patches to low branches and called sweetly to me as they enticed me far down the roads and skidding trails. When the trails ended, they followed me back to the skidways, continually calling and fluttering nearby.

On another afternoon I heard the loud hammering of a bird, "chock, chock, chock!" on a hollow tree ahead of me down the road. Making Tex stay beside me, I approached as quietly as I could. Off to one side and only a few feet from the road, I discovered the hollow elm with a magnificent bird clinging to the tree halfway up and digging expertly for grubs, I suppose, or bugs in the bark. He looked and acted like an overgrown woodpecker. His body was of the same bright coloring. His long bill, his brilliant red crest and black-and-white-striped head and thin neck were like a woodpecker's bill, crest, and head. He was hammering with greedy unconcern and did not see me until I stood about twelve feet from the foot of the tree.

When he did discover my presence, he merely swung around out of my sight on the other side of the hollow tree and farther up. I picked my way closer, off the road and into the deep snow—still trying to keep Tex at my heels, and keep him quiet! My second glimpse of large bright-red crest and long yellow beak was fleeting, for the bird quickly climbed up and around the tree. Then he again began his hammering. Again I went around the tree, and went closer, thinking he might become accustomed to me. But the bird eluded me again. All I saw this time was his gay red crest and a beady eye peering at me from a newly assumed position. Deep snow covering branches over a hole made a perfect trap for curious me, and I walked into it while I gazed upward. I crashed to the bottom of the hole, and branches and snow covered me.

Tex ran around the edge of the hollow and barked. The bird—I later learned from Howard it was a pilated woodpecker—that had played hide-and-seek with me, now swooped low over my head, darted at Tex, circled into the sky, and flew away laughing to himself. He laughed, I am sure, at such a foolish creature as I was, trapped and tangled in branches and sprinkled with wet, cold snow. This inconvenience was worth the view I had of him as he spread his black wings and exposed their white lining. I crawled out of my trap, and Tex pursued an arrogant woodpecker—which must have led him a merry chase, because he came home that evening too tired to play.

I heard hammering for several days around the woods roads, but I never saw this woodpecker again. I wished for just one long look at him, because he was such a cocky, handsome fellow.

With little effort, except for sweeping and straightening the van counter, I kept the office and our bedroom clean. The bull cook, whose job it was to carry wood and water, build fires, clean and fill lamps, and sweep the floors, mopped the floor with hot soapy water every Saturday after he had carried out the ashes and filled the water bucket and the wood rack. Bookkeeping, sketching, and reading occupied a great deal of my time, but the activity in the woods lured me out to the timber for a part of nearly every day.

I spent many afternoons walking—with an occasional rest on some stump at a safe distance from the loading crew. If they were loading veneer, Mr. Rogers amiably chatted with me. He told me of the first days of the veneer mill at Mellen, of the struggle to make contacts with lasting markets, and of the company's final success in getting established. The enterprise of establishing this veneer mill in the heart of the hardwood country contributed on a large scale to the effort of winning World War II, as did the magnificent efforts of all other mills. Howard was glad to have a market close by for our birch and maple veneer.

When Mr. Rogers was not around for me to talk to, there

was always the fascination of watching the teamsters pile the logs up on the skidways—just to have the loading crew snatch them away and swing them up onto the trucks. Sometimes the tractor was used on the crosshaul, but the fast-moving, hard-pulling horses made loading more of an excitement. Their harnesses jangled; the teamster, on the way back, ran to keep up with the teams.

Loading logs and hauling them as soon as they are cut is called "hot logging." Most of the logging Howard did at Tula Camp was hot logging, but before the end of the first winter he found it necessary to clear places along the woods roads where he decked nearly half a million feet of logs. An impressive sight they were—these immense decks covered with snow and waiting to be hauled.

Often as I watched Ray and Clarence quickly load one truck after another, I thought the hooker's job looked very easy. And I wondered why Howard chose his hookers with so much care. Then one afternoon I looked on keenly while the loading crew showed a greenhorn how hooking logs is done. A forest ranger from an office in Lansing had come to Tula Camp to discuss selective logging which Howard had agreed to do for the state in a piece of timber adjoining ours. This official stopped at the skidway where Ray and Clarence were loading veneer while Howard went in on a sawyers' strip. The forester watched the crew swing the logs up on to the load with dexterity and ease. He made casual remarks.

"That's nice veneer. Where's it being sent?"

Mr. Rogers told him about the Mellen veneer mill and said, "Best veneer we can buy is coming out o' these woods."

Then the forester made a grave mistake by kidding the hookers. "You fellows have it pretty soft, hanging onto those ropes all day."

Ray and Clarence, who had never seen the fellow before, began to work on him by telling him, "Sure, this ain't nothin'."

"Hot Logging" Veneer

"Better'n being inside all day," Clarence said.

" 'Tain't much work; just gotta be here in case—say, why don't ya try 'er yourself and see how easy it is?" offered Ray.

"Sure, go on. Try 'er. It's easy as fallin' off a log." Clarence's convincing tone of voice and his own easy motions had the desired effect on the young forester.

Clarence handed him the limp and seemingly harmless loading rope.

The easily persuaded greenhorn stepped right up and grabbed the sucker line (that could be a pun) and the grab hook which the professionals call a "pup." Clarence carefully showed him exactly how to place the pup on the end of the log. Even in good fun, he did not want the fellow to lose a finger or two when the loading cable tightened and the pup grabbed and dug into the end of the log. With sly winks from the hookers to the crosshaul teamster and the loader, Clarence put the greenhorn "on his own."

Clarence whispered to us, "When a stranger in the woods begins makin' big talk about how easy he *thinks* it is just t' hang onta a sucker line all day, *that* greenhorn better get ready t' be 'shown the ropes.' "

Mr. Rogers folded his arms across his chest and planted his feet apart to brace himself, I suppose, for what he knew was coming. I watched closely, not knowing what to expect.

The first and second logs rolled up onto the truck with ease and no apparent effort on the part of the forester. The grin on the teamster's face widened, and the top loader climbed down to a safe place on the ground. Ray pulled his cap low over his eyes to hide his amusement and took a much firmer grip on his sucker line.

Then things began to happen fast.

The next log bounced over another small log and jerked the forester hard. The log went wild as it ascended the skids and swung toward him. The forester ducked his head quickly. He could not control his end of the log, although he hung onto the line and fought the sudden antics of the log until (by

Ray's clever handling of the other sucker line) it banged down hard on the top of the load. Watching out of the corner of his eye, Clarence enjoyed the show and smoked a cigarette.

I understood now that the forester was in for a rough time.

Mr. Rogers and I moved far back and tried hard to keep from laughing. We should have hated to spoil the hookers' fun. The next log behaved just as crazily as the first, and sent the citified forester running all the way around the back of the truck where he landed sprawling in the brush.

He brushed off the twigs and snow, unaware of how he was being played for a sucker.

The fast trip to the brush should have been the climax. But the eager forester still came back for more, only to find that a swinging log could nearly tear his arms from his body.

Once, in his effort to keep a big log from bumping the side of the cab, he hit the ground on his seat. As the sucker line and big log dragged him across the skidway, a jagged root ripped a big three-cornered tear in his pants.

Now I knew that hooking can be easy, but never for a man who knows all about it before he has tried to hook at least one load with the co-operation of the whole crew.

No wonder Howard hired amiable, husky, willing men to swamp, hook, and drive trucks.

Howard returned from the sawyers' strip and took in the whole situation at a glance. "Guess the crew took a little press out of your pants while I was gone. It's getting late. We better get out to that timber."

The last were kind words to the forest ranger, whose body and pride were quite obviously bruised and beaten. He neither smiled nor spoke as he dropped the sucker line and climbed into Howard's car.

When they were out of sight, the hookers got down to business, reloading the logs that crisscrossed the top of the load. There was a keen edge to their laughter and their jokes.

Mr. Rogers turned to me with his eyes crinkled in a big tender smile. "The starch is oot o' the lad now. He'll wear the better for it." The preceding words are my favorite line of Mr. Rogers'.

"I wonder what alibi he will have to explain the condition of his clothes," I said, surmising that the true story would not be heard from him, except by his wide-eyed grandchildren.

The entire crew heard before suppertime the story of a wise guy who learned the ropes. They laughed about it for weeks. To see or hear about a man work so hard doing what Ray and Clarence did with ease was funny. The men would not have been lumberjacks if they had not taken advantage of such a perfect opportunity to enliven the hard day's work with some good fun. When I told Howard about the forester's hooking, he said, "Boy, oh boy! If I'd been there he'd had a still tougher time hanging on to that rope. The guys were too easy."

One day I followed a skidding trail in to Ole's and Emil's strip. I wanted to watch them saw.

Low cedar branches laden with snow slapped me in the face. Cedar waxwings hidden in the high branches of the tall hemlocks whistled their high, thin call as I approached their sanctuary, and the finches answered melodiously with their silvery notes.

When I heard chopping I called, "Hello-o, Ole," to let the sawyers know I was nearby.

The chopping stopped, and I called again.

"Over here ve be," Emil called to direct me to their spot.

When I found them they greeted me, indicated by jerky thumbs a safe stump from which I could watch, and went quickly back to their felling.

Together they chopped out a wedge-shaped chunk from the trunk of the tree they were going to fell. With each alternate, perfectly timed stroke from first Ole's ax and then Emil's blade, a butter-colored chip curved out into the snow until the notching was finished.

They propped their axes out of the way, found firm footing

on the opposite side of the tree from where they had cut the notch, and began drawing their crosscut back and forth, pausing now and then and going back to their sawing with even, easy movements. Emil pounded two hardwood wedges in behind their saw blade—the whack of the flat side of the ax on the hardwood sounding rhythmically through the woods —and they resumed their final cutting.

Tex came out of the trail, sniffed my tracks, and bounded over brush to greet me. Then he crouched behind the sawyers, with first one ear and then the other cocked to hear the sound of the first crack of the tree. When Emil and Ole drew their saw out of the cut and backed away, Tex pointed both ears and quivered with anticipation.

"Tim-ber-r-r!" Ole called.

He glanced in my direction and then watched the treetop tremble.

The big maple leaned. Slowly, slowly, and then faster, faster, it came. With crashing and cracking of branches, it hit the ground.

The trunk rolled away from the stump surrounded by chips and sawdust, and Tex sprang for the quivering leafless branches to shake them and growl.

The sawyers bucked off the big limbs and branches.

Emil laid their measuring pole along the tree.

Ole notched the bark at the measured intervals.

Together they chopped off branches and brush interfering with the cuts they were going to make.

Now they began sawing the tree into log lengths.

The fragrance of fresh-cut hardwood, stronger than the smell of new lumber, came to me on a cool breeze. There came, too, the odor of the kerosene Emil poured on their saw to prevent it from becoming pinched in the log cuts. The winter sunshine warmed my shoulders.

The logs were sawed, Emil proceeded to mark the stump with the total number of these.

The pair of them, Ole and Emil, chose the next tree to come down.

"Better ve get dat schoolmarm out of the vay of dat burly ol' birch first," Emil advised Ole.

I thought they meant that I was in the way, and I asked, "Where will I be safe?"

"You not be in vay, Missus. Yust dat big twisted maple vit da vide crotch. Ve say dat's a 'schoolmarm.' You be safe. Ve no vidow-maker you."

Emil and Ole cut out interfering brush and saplings and began felling the schoolmarm as I left the woods.

All day long this pair of old-timers sent trees crashing to the ground. Sometimes a tree would unexpectedly lean and get hung up in another, but as a rule nothing kept these men from great accuracy. Rare was the tree they felled that did not lie prostrate and outstretched within a foot of where they had planned.

Because of the extreme danger of staying where the sawyers were cutting, even when they knew I was watching, I spent more time walking and watching the loading crews.

In the late afternoon sunshine, that particular day, I retraced my steps along the skidding trail and rode back to camp with the truck driver, Joe.

We drove up to the office, and Joe stopped for gas and coffee.

My logger's camp! It was a haven to me. In the deep snow its buildings nestled homely together. Good-smelling wood smoke slowly curled upwards from the stovepipe chimneys. On the sunny sides of the buildings, the even drips from the icicles sparkled at rhythmic intervals, some at a fast tempo and others more slowly. My rosy-breasted birds called to me from the saplings. . . .

I followed Joe into the cookshanty to have coffee from the big black pot and also find some crumbs for the finches.

As I walked to camp on another day, I heard the South Shore engine whistle, first distressingly long and drawn out, then excitedly staccato. The whistling continued, and I wondered why. At that moment Howard came speeding from the woods in the car. He slid to a stop, and called to me, "I'm going out to the railroad. Seems like something is wrong. Want to come along?"

"Sure!" I climbed in. We drove through camp so fast that Howard failed to see the cook signaling to us for a ride. Fortunately for us there were no log trucks coming back empty, or we might never have arrived at the railroad crossing.

The jerky whistling sounded again. "What do you think is wrong?" I shouted. But Howard was too busy rounding curves to answer.

As we approached Watermelon's place we saw that the gate was open! Only a thin wisp of smoke curled from the chimney. The house appeared to be deserted. We sensed what was wrong and drove on to the crossing.

The coal car blocked the road. We parked the car and walked down to the engine. The engineer was climbing out of the cab, Howard called to him, "What's the trouble?"

The engineer did not hear Howard. We followed him. We walked to the front of the engine, and there, in the middle of the tracks and staunchly defying the train or the engineer to come closer, stood one of Watermelon's cows! Firmly she maintained her primordial right. The engineer flailed his arms and yelled at her; but the closer he came to her, the less co-operative she became. He coaxed her, then threatened her with his clenched fist. He tugged at her neck, and she still refused to move an inch. Howard helped by tugging and threatening her, too, while I watched. Their combined efforts met with no success.

I looked on in silence as long as I could, then called out a suggestion. "Twist her tail, and she'll move fast."

Howard gave me a disgusted look, as if to say, "What a silly idea," but the exasperated engineer reached behind the cow, grabbed her tail, and twisted it with vengeance. The cow kicked up her heels, mooed, and loped off the tracks.

"Be danged if it didn't work," commented the amazed engineer to Howard, who was also dumfounded at the stubborn cow's quick reaction. As the engineer climbed back in his cab he called a grateful, "Thanks, lady!"

I chased the cow back through the gate while Howard turned the car around. As I closed the gate I heard the train start on its way.

That night in the office when Howard asked, "Who the heck left Watermelon's gate open?" all the truck drivers passed the buck. Every man claimed he hadn't been so thoughtless. But the gate was never left open again.

As I left the cookshanty, one noon, Florida called to me, "Tell Howard there's some pie and doughnuts to take as a treat to Arvid today."

"If I can get Cowboy to lend me his sled, I'll take them out to him. I'm going to walk down that way."

I found Cowboy splitting wood by the bunkhouse. Together we piled the old magazines from the bunkhouse and cookshanty on the sled he used for hauling wood. The cook put the pie and doughnuts in a box and tied it securely on the sled. I raced down the hills with the sled bumping my ankles and puffed up the inclines with the heavy load dragging far behind me. Tex barked loudly and nipped at the sled when he was not bounding off into the brush after birds or squirrels.

Arvid, a Swede piecemaker, had built his shack beyond the hauling roads where the timber was best for cutting short stuff. Before I reached his clearing the sled became increasingly difficult to drag up even the smallest hills. But the sight of his tiny black shack between the trees restored my energy. Because recent snowfalls had covered any path there might have been from the road to his shack, I left the sled on the road and tramped through the unbroken deep snow to find Arvid. As I made my way around trees and over snow-covered logs I called, "Hello, Arvid." My small voice in this silent forest sounded like the voice of a giant.

Arvid, leaning on a cane, came to the doorway of his shack. "Hello, Missus. Vat you be doing here?"

'I've brought you some magazines, a pie, and some fresh doughnuts. I left them on the sled in the road." Arvid's eyes brightened, and a smile wrinkled his leather-brown face.

Tex wagged his tail in recognition of an old friend, then went off on a hunting excursion behind the shack. I made my way down a trail to Arvid's door that was flanked on one side by a bench and an overturned washtub and on the other by a grindstone, axes, and a queer sled fashioned much like an Eskimo dog sled with waisthigh handles at the back.

"Pie and magazines you say? Dat's goot. Ve'll load dem on my sled. Easier tu get dem here dat vay, yah?"

As he spoke he cumbersomely made his way to his sled and began breaking a trail back to the road by pushing his sled ahead of him. I followed. At the road we transferred the magazines and box to the platform on Arvid's sled. He lifted the cover of the box, and the odor of fresh-baked apple pie steamed out into the cold.

"Ah, apple pie. I sure like apple pie. Vonce I make apple pie, but I'm no baker. Dat pie vas not so goot. I vas a carpenter in Sveeden and had no time tu learn baking."

"How long have you been in this country, Arvid?"

"In nineteen-vun I come tu Nu York. I vent tu Canada und vorked in da voods for a vile. Den I come tu Michigan und cut vite pine."

"What made you start shacking?"

"Vel, it vas dis vay, Missus. A log, tundering down from a rollway, broke da leg I broke t'vice in Sveeden ven I vas a boy. I yust missed getting killt by da big log und vas glad tu have only annuder busted leg. I never got around goot after dat, und I couldn't find no saw partner. So—I vent tu shacking." He pushed his sled around so that we could unload it easily.

"Did you make your own sled?"

"Yah, Missus. Und vonce I made a vagon in Sveeden, veels und all," he told me proudly.

While Arvid unloaded the magazines, a few at a time, and stacked them beside his battered rocker, I carried the box of bakery in and left it on the table amid stacks of clean dishes and tableware. The pale winter sunlight slanting through the only window disclosed two up-ended orange crates filled with boxes, packages, and cans of food. These crates served also as a stand for a granite washbasin and water bucket. Above these hung a grimy towel beside a kerosene lamp in a wall bracket. In the gloom beyond the window stood a gray-blanketed bed. From nails driven in the walls above the bed hung a lantern and clothes. A pot of coffee simmered on the small black cookstove. The smell of the coffee along with that of sweaty clothes, cooking, old newspapers, stove wood, soap, and kerosene produced a peculiar musty smell that hung thick in the poorly ventilated shack.

Arvid straightened up after sliding the last of the magazines onto the pile and motioned to the black pot on the stove. "Vee have a cup 'a coffee vile I get ready my supper, yah?" He poured the coffee and offered me sugar and canned milk to cool it. He selected several sticks of wood from a pile in the corner and filled his stove. Then he punched holes in the top of a can of beans and slid it into the oven beside a small covered pan which I supposed contained meat. From a box of potatoes he selected a large one which he rolled in beside the beans. With the aid of the handle of his cane, he pulled the oven door up and closed it.

"Now, Missus, vile I vork my supper bake." He stirred his coffee vigorously and drank it all in several swallows. From a hook by the door he jerked his jacket and cap. As we went outside he slid into the heavy plaid coat and slapped his cap over his untrimmed gray hair.

"I bain cutting ties dis month. Vant tu come along?"

"If I won't bother you I'd like to see how you manage to pile those heavy ties and posts.

"No bother, Missus. I do it vit my sled easy. See."

He selected four peeled ties, rolled them from the bark-covered snow onto the low platform of his sled; and, gripping the sled handles, he pushed the ties to a pile already started under the snow-covered boughs of a big hemlock. Finding firm footing, he flipped the ties off the sled and rolled them onto the pile. However, if he had not once been a stalwart man with strong arms he could never have done that work alone.

"How do you get around in the woods when the snow is gone and you can't use your sled for support, Arvid?"

"Den my cane is enough. Da ground's not slippy den."

Arvid, preparing to go farther into the woods to cut more ties, laid his saw and ax on his sled. I thought of the long walk back to camp.

"I'd better start back now Arvid, but I'll come again soon with more pies."

From his shirt pocket Arvid pulled a slip of tablet paper. "Yu tell Howard tu get dese tings for next time. Und tanks for yu visit, Missus."

I tucked the list safely in my pocket and whistled to Tex. As I passed Arvid's tiny shack I thought of his lonely life, with not even a dog to whom he could give a friendly pat. I turned to wave.

He called, "Goot-bye, Missus."

The first shacker I had encountered before Howard and I were married appeared clad only in dirty underwear and waved a whisky bottle. Unlike that specimen, Arvid dressed warmly and apparently enjoyed coffee more than whisky. And, judging from his orderly shack, he was a thrifty, sensible shacker. He was one of a great number of recluses whom city people seldom see, who work every day sawing and hewing ties or lagging, who keep their own records of forest products they cut and pile, and whose accounts are settled once a year by the logger who employs them. Their tarpapered cabins dot the timbered areas as well as the cut-over regions of the Upper Peninsula.

I rode the sled down the hills back to camp. But it was hard to steer, and my rides frequently ended in snowbanks. Tex, barking and tugging at my sleeve, ran beside me. Climbing the last hills left us puffing and exhausted. I was glad to return the sled to Cowboy, whose pile of split wood had grown to an enormous size while I had been gone.

The smell of baking bread coming from the cookshanty made me hungry. While I tried to resist the temptation of fresh bread, two State Police, clad in trim bright blue uniforms and shining black puttees, came out into the sunshine and held open the door. "Thanks for the coffee, Mrs. Anderson. Tell Howard we stopped in."

"I'll do that. Stop again sometime." These police were frequent visitors at Tula Camp and personal friends of Howard's —welcome contacts with life outside of camp.

The good smell coming from the cookshanty overwhelmed me. I went in for a lunch of strong camp coffee and hot bread spread with thick plum jam. Any time after the noon dishes had been washed and the cookee had set the tables until the cook put the potatoes on the stove to boil for supper was lunchtime at Tula Camp. Everyone who came to camp stopped for coffee and whatever else was on the lunch table—doughnuts, pie, cake, cookies, cold pancakes, cold chops, or crackers and cheese. The big black coffeepot at the back of the stove never stood empty or cold. When Florida, the cook, looked at the clock and dumped the spuds in the kettle, I climbed the path to the office to wait for the return of my logger husband and the call to supper.

Sweet Taste of Supper

Once a week the cook's supplies needed to be replenished. Although I had never driven much, Howard trusted me to be careful, and I drove to Marenisco for groceries enough times to qualify for a driver's license. Along with the license, I acquired the regular job of toting in supplies for Tula Camp. On stormy days as well as in clear sunny weather, I started after dinner to drive to Marenisco. In the glove compartment I stowed away lists of things to buy—groceries and meat, office supplies, blacksmith things, and a list of odds and ends to collect like thread, bandages, colic medicine, a dozen mousetraps, and a special kind of plug tobacco for Emil. On the highway the truck drivers waved to me. I returned their greetings, hoping that if I ever got in a jam they would stop and help me. I'm sure they would have, too, but, fortunately for me, I never needed their help. I always arrived safely at the store and invariably back to Tula Camp.

Bonifas General Store at Marenisco always buzzed with the hum of conversation between mill men's wives, lumberjacks, and genial clerks. It would characteristically be interrupted at regular intervals by the ring of the cash register and shrill blasts of the mill whistle. Logging trucks ground past the front entrance in continual parade on the way to the mill yard behind the store. Sawdust tracked the floor of the store, and the combined smells of groceries, fresh and smoked meats, sacks of flour and feed, harnesses, and fuel oil made this a typical mill-town store. When I went in I would hand my lists to the busiest clerk, and within an hour the wants of Tula Camp would be piled on the back seat and tightly packed in the trunk of the car. (We later bought a pickup for hauling supplies to camp.)

On my way back I would drive in to Wakefield and stop for our mail. Oh, happy was my day when a bit of mail came that I did not have to enter in the account books. Sometimes I stopped at our apartment to pick up our laundry and leave the things to be sent out again. On other days I took a bath, watered my plants, and replenished my dwindling sketching material or picked up a new book. These hours alone, spent in the silence and staleness of our apartment which we occupied only on Sundays, would send me scurrying back to Howard and to the activity at Tula Camp in time for supper.

Only once was I too late to eat with Howard. That was because of unusual conditions over which I had little control. As I drove up to Watermelon's gate late one afternoon, he came running down the hill. His shirttail flapped behind him, and he yelled loudly in Italian. Under one arm he clutched a squawking chicken. Under the other arm he hugged a squirming, squealing pink pig. (He looked so funny I later made a sketch of him.)

I slammed on the brakes and stopped!

When Watermelon could talk he panted, "Here's—nice chicken stew—for Mamma. Dis-a hen—she no lay-a—da eggs. No can feed da lazy hen. You make-a da good chicken stew, yes?"

"Yes, I can make good chicken stew, Watermelon."

"Here, den." Watermelon shifted his pig to another position between his knees. He held the squirming animal there while he opened the car door and dumped the chicken on the groceries. The chicken's legs were tied, but she flapped her wings and squawked and squawked.

Watermelon lifted the pig into his arms again and grinned. "Make-a da good stew, dat hen."

I thanked him for the chicken and was glad to see that he intended to keep his squealing pig. Watermelon opened the gate and waved. I started on to camp with the irritated hen irritating me. The smell of the onions in the back seat combined with the chicken smell made me feel sick. I rolled the window down for fresh air. A cool breeze fanned my hot face. The last of the winter sun rapidly disappeared behind the black hemlocks. Before I reached the field complete darkness obscured the familiar road, and I switched on the headlights.

The fresh air revived me. The chicken stopped squawking. But I should have known that her silence meant only that she had freed herself of the cord around her feet. With a determined cackle and a flutter of feathers, she perched herself on my shoulder. Surprised and startled, I headed for the snowbank but swerved back in time to avoid getting stuck.

That evening I arrived at camp late for supper. Only Ed the foreman saw me drive in with a ruffled white chicken perched cozily on my shoulder. Obligingly, Ed caught her, tied her legs, and took her to the cook, who killed and plucked her for us. "Ole lazy hen dat no lay-a da eggs" came out of my gently simmering pot, sweet and tender for our Sunday dinner. But even so, I never forgave her for making me late for supper with Howard.

Late in the afternoon, when great numbers of fresh-cut logs lay inert along the skidding trails waiting for the clasp of the skidding tongs, tired sawyers pulled their saws slowly out of their last cuts. Sensing that it was quitting time, perhaps by the long shadows moving across the sawdust-covered snow, or perhaps by the ache in his back and arms, Emil would pull his watch out of his pocket to verify his hunch that the day's work was done.

"Guess ve go camp now, Ole," Emil would say as he slid his watch in his pocket. Ole would lean his ax on the protected side of a stump and lay the saw safely on a log nearby. They would pick up their jackets from a log and start down the trail.

When they appeared on the skidway jammed with logs ready for loading in the morning, Ed would look at his watch and yell out for all around to hear, "Call it a day, men!"

On the days when the skidways were only partly filled, Ed kept the teamsters at their job of dragging out logs a while longer by saying, "Better yank out a few birch on the veneer skidway, then we'll be all set to load three trucks first thing in the morning."

The teamsters would urge their horses into a fast gait with a flick of the reins, and in a short time logs choked the veneer skidway. The jammer stood spotted for an early start in the morning. Hookers secured the guy lines to sturdy saplings and grabbed their jackets from a stump. Tired as they were, they would walk back to camp at a fast pace. They were joined by Emil and Ole and other sawyers swinging empty kerosene jugs. Some sawyers came out from the skidding trails with saws that needed to be sharpened swung over their shoulders. The limber gray-blue blades bent rhythmically to the jacks' long strides. The tired men spoke little as they climbed the hills back to camp. Tex walked calmly by first one gang of sawyers and then by another.

Once the teams sensed that they were headed for the barn, the teamsters were forced to run to keep close behind them. Steve, the fiery Frenchman, would fasten the loose ends of his reins up on the strap across King's flank, unclamp the evener, and prop it by the jammer. With a slap on Paddy's shoulder, Steve would send his team down the road alone. If Howard expected to join me at the skidway for the walk to camp, he left word with one of the men. Usually Steve or Ed knew where Howard was. "Howard said to tell ya he'd be late," Steve would call back as he followed Paddy and King down the road. "Had ta blaze some strips for a couple a new saw gangs. Said fer ya not ta wait, but ta ride with

Charlie." Then I would climb up beside Charlie and ride on the roaring tractor into camp.

On days when I waited at camp for quitting time, Charlies' tractor, as it sounded louder and louder, told me that the crew was coming in. To the cook the roar of the tractor meant that hungry jacks soon would stream into camp for him to feed, and that it was time to build up the fire and trim the steaks. As I watched from the south window in the office, Charlie, usually the first man back to camp, would swing the tractor around in front of the tractor shed and let the motor idle while he cleaned out with a long iron bar the icy snow clinging in chunks to tracks. While he greased the tractor and filled the fuel tank, the teams, their flanks steaming and their nostrils rimmed with frost, would jog down to the watering trough behind the bunkhouse.

Jacks with frosty eyebrows and whiskers would straggle in to camp behind the teams. As they called to Paul when they passed his shop, their breath steamed out in white puffs. Men's husky voices, the purr of the tractor motor, and the jangle of harnesses broke the afternoon quietness of camp. Blue-gray wisps of wood smoke curled out of the stovepipe chinmeys to scent the frosty air.

Truck drivers would return from their last trips to Brandt's Landing or Bonifas Mill with empty trailers, which they deftly backed into their parking places. After they brought their scale sheets to the office they joined the lumberjacks washing up in bunkhouse.

Ed would come into the office, jerk off his mitts and cap, and without taking off his jacket he would warm his hands, his front, and his back at the fire crackling in the stove. "Another day, another dollar," he'd say, and pour hot water from the teakettle into the washbasin.

The hot water he splashed on his rough and red face melted the icy coating on his eyebrows and eased the ache in his stiff, cold hands. In the fast fading light of the late afternoon sun he combed his curly graying hair before the small mirror. (Waiting their turn with tired patience, I'm sure, the jacks washed up in the bunkhouse in much the same manner as Ed did.) From under his bunk Ed would then pull his low-cut rubbers, which he wore in the evenings around the office to warm and rest his cold, tired feet. The springs of his bunk always squeaked as he flung himself down to have a quick nap before the supper bell. Occasionally a teamster would rouse him for a few minutes to tell him of Captain's sore shoulder or of a broken evener to be repaired. The unmistakable odor of the barn and horses the teamster brought with him would settle thickly over everything. Then the smell would be quickly diluted with the fresh cold draft the teamster let in as he, too, went to wait his turn to wash up and comb his hair.

Howard usually arrived from the woods last of all the men. As he warmed his cold hands on my flushed cheeks he would give me a frosty kiss. "What's for supper? I'm starved." The strong day's work kept Howard ravenously hungry.

"I saw pies cooling and a cake waiting to be frosted when I went for coffee. But the meat is special. You'll be surprised."

"I'll bet we're having steak."

"Sure thing. You guessed right."

Howard would splash soapsuds and water with vigor, shave off his red stubble of whiskers, and brush back his hair into the soft wave that made him look so youthful. I would change my warm shirt to a bright plaid blouse, comb my hair, and dab on a bit of gay lipstick.

The melodious ring of the triangle would call Ed and the lumberjacks to supper. Capless and coatless men left the bunkhouse in a solemn stream, great fatigue showing plainly on their faces and in their dragging steps. They filed into the cookshanty singly. As we watched them from the office window, their dark figures against the last radiant glow in the western sky cast Paul Bunyan shadows across the road. The brilliant winter sunsets, the giant purple shadows of tired jacks, and the bell-like ring of the triangle remain symbolic in my mind of the end of a day's work for lumberjacks.

When the office door slammed behind Ed, Howard would pull me into his arms and give me a strong hug. "I'm glad I don't have to drive to Wakefield every night for supper after a day in the woods. It's grand of you to stay out here with only the jacks and me for company."

"But I love it. In town alone I'd be lonesome all day. I get lots done that I couldn't do if I had housework and cooking to do."

"Sure you get lots done—bookkeeping that you hate."

"I don't mind the bookkeeping, really. I just hope I don't have to use red ink."

"Looks like you won't have to use red ink this year. Things are going good. Just *hope* that this nice weather holds."

Together we would light the lamps. Darkness would close around our tiny shack. With the sun gone, a chill turning colder by the minute would settle over Tula Camp. When Ed came back from eating his supper, he would fill the stove carefully with big chunks of long-burning hardwood.

"Better put on a jacket when you go to eat," Ed would advise us. "The temperature's gonna drop 'way down below zero tonight. In a doggone hurry, too."

We would step out into a black starlit night. Through the smoky windows of the bunkhouse the lighted lamps sent frail shafts of transparent gold onto the shimmering snow. On our way to the cookshanty I could glimpse shadowy figures of jacks bending toward the radio by the window, socks steaming above the stove's belly that glowed with a reddish cast, and tired jacks climbing into their bunks for the night.

"Do they all go to bed so soon after supper, Howard?"

"Oh, the sports play poker a while; the lonely read or swap stories. The cold huddle by the stove, while the 'stake-hungry' file saws or mend their boots. The tired just snore."

Sonorus Song—the Tired Snored

Howard and I ate only this one meal together. Sometimes we were joined by truck drivers coming in late from their last haul. Then we dropped the no-talking rule, and I came to know some of the young fellows who drove the trucks. We would laugh and joke and enjoy the tasty, hot food, of which there was always enough to feed ten or fifteen more people. The smell of spicy pumpkin pie and cinnamon rolls tempted us to linger in the cookshanty. But the cook, the first one up in the morning, was likewise the first one to bed. We would have been thoughtless to keep him up long after supper. As soon as the cookee had done the dishes and set the tables for breakfast the cook blew out his lamps and carried his alarm clock with him behind the curtain of his cubicle. We would go back to our office, enjoying the short walk alone in the still cold winter night.

Eating supper early made the winter evenings long. With nothing much to do, a few of the least tired jacks would wander into the office for tobacco and candy bars, to talk a little, to kid each other and Howard, and to play a game of blackjack with me. Howard always kept a punch board at which they tried their luck. Jacks, gamblers at heart, like most loggers, enjoyed taking a chance or two. Blackjack and twenty-one aces were the games I enjoyed most. When the jacks went back to the bunkhouse, Howard and I would play a game of cribbage. Ed checked strip counts and then would call it a day. At eight o' clock we would turn out the lamps, call Tex in, and hit the hay, too.

Around Our Barrel Stove

Some evenings, while Howard balanced himself on the golden-yellow three-legged stool his dad had made from a crotched maple limb, and while Baldy leaned on the counter, lonesome jacks sat on Ed's bunk and swapped stories with Howard. I curled my cold feet under me on the blanketed car cushion fitted to the top of a long shipping box (thus our davenport) to listen to their stories about other loggers and early logging days. The lamp smoked and the fire sputtered and burned low from lack of attention. The men became wholly engrossed in storytelling. I became completely absorbed in listening.

"If you wasn't so lazy, Howard, you'd be knittin' like your dad used to do when I worked for him over by Ethelwood Landing." By the twinkle in Baldy's eyes I knew he enjoyed kidding Howard.

"I didn't know your father could knit, Howard," Ed said.

"Sure. Dad was an expert knitter. He learned to knit so he wouldn't have to help with the outside chores on the farm. While his brothers and sisters gathered the eggs and milked the cows, he sat by the warm kitchen stove and knitted stockings for all the family. When Dad's uncle, Lewis Jenson, offered to send him to business college, he was glad of the chance to leave the farm."

"Funny thing, Jenson's comin' up here when he could'a made a mint more o' money by goin' West with the rest of the Michigan lumbermen. Guess Jenson was worth prit' nigh a million bucks when he started his mill at Ewen. Ain't thet so, Howard?"

"People said so; and I guess he was, at that."

"And I suppose he was a rounder, too, like all timber barons," I surmised.

"No, not all were the rounders you think, Julie. Mr. Jenson was a highly respected lumberman in the Upper Peninsula—not because he flaunted his wealth, but because he gave financial aid to so many people and organizations."

"His place was that one in Ewen with the rows of elms on the front lawn, weren't it?" Ed asked. "I heard it said he planted them trees."

"Yes, that's his old place. He liked farming and growing things just about as much as he liked lumbering, I guess. He stayed active long after he quit lumbering. True enough, as my dad said, he had to stop on the trail every ten steps to shake the numbness out of his left leg in order to keep up with other men. But that didn't keep him from looking timber as he had done so often in the old days. To the timber he looked he drove a buckboard. But to the three farms he owned, he rode a lumber wagon, seating himself on the reach behind his single horse. For comfortable riding he always used a 'dry ass.'"

Howard turned to me and explained, "That's nothing more than a gunny sack filled with straw. Lumberjack teamsters sat on it when they rode a load of logs on the sleigh haul. The dry ass protected them from the cold and wet of the icy logs or the bunk of the sleigh on the return trip. For an oldish man—and Mr. Jenson was well up in years when I knew him—it eased the rough riding over country roads in all seasons."

I thought how truthful and blunt the language of the lumberjacks was. But I laughed out loud when I imagined one lumberjack yelling to another, "Toss me that comforting cushion, will ya, Joe?" Jacks were (and still are) husky men. They used husky words.

"Men like Jenson had a good chance to make millions of bucks. Timber was cheap—only a dollar and twenty-five cents an acre—darn cheap. Jacks were easy to get darn cheap, too," Ed said wistfully. "Michigan even *gave away* section after section of timber, I'm told."

"Yes, this piece of timber I'm logging is part of an original land grant from the state," Howard said. "When Michigan was poor in cash but wealthy in pine, in order to get a canal built through the Keweenaw Peninsula, a land deal gave the Keweenaw Land Association acres of timbered regions up here for building the canal. In the same way, the railroad companies were given timber—every odd section along their railroads—just for building the railroads into this region. If these sections could be colored on a map it would look like a checkerboard strip."

Just then Steve came in from the barn. The smell of liniment reeked around him. "Been fixin' up King's sore shoulder. I'm gonna need some more juniper berries. Better get some, 'cause I don't think you got any more." He glanced in my direction and then told the men, "Say there's an ole hound hanging 'round down by the barn."

"What kind of a hound, Steve?" Baldy sounded overcurious but that meant nothing to me then.

"Ax-handle hound, he is, from the look at 'im I got."

"Ax-handle hound," I said in amazement. "What in the world kind of a dog is that?"

Steve's eyes twinkled.

"That's the kind of dog that prowls around all camps looking for peavy handles to chew."

While everybody laughed long and loudly, Howard explained, "Ax-handle hound is a jack's name for a porcupine, Julie."

Someone mentioned bird's-eye maple. "Funny, ain't it,"

said Steve, "that those trees should be so scarce and bring such big money? Someone ought ta raise a few maple, put some choice bugs on 'em and let the birds go to it and peck away making bird's-eye maple."

"Bird's-eye maple. Is that—?" I asked before I thought.

Steve took up the story quickly. "Sure, hain't you never heard about bird's-eye maple? It's like this: a nice tall maple grows along to be a big tree. Then it gits covered with bugs, certain kinda bugs, big ones, choice eatin' for the birds. The birds discover the bugs, and a flock of 'em comes and set ta peckin' out the bugs from the bark and under it. These birds, sorta woodpeckers like, have long bills that go into the tree a little way and leave a scar. It's them peck marks that makes the bird's-eye in the maple trees." A sly grin crept over Steve's mouth, and his black eyes sparkled with amusement at my honest surprise.

All I could say was, "Well, I'll be darned." Now I would say, "Oh, yes?" Steve had hoodwinked me with his fantastic tale which was almost believable until Howard told me that no one is sure just what does cause the eyes in bird's-eye maple.

The conversation drifted to the nameless roaring, cursing jacks who slashed away at the impressive stands of pine and logged like madmen all winter, making a stake and raising a "helluva" thirst for liquor and women. Skilled and daring birlers who rolled the logs down the rivers and rode the foam when the logs thinned out were remembered. Tales of how the jacks were welcomed by whorish maids, how they spent their money on booze, and how they engaged in magnificent brawls were told by jacks tongue-tied until they thought I wasn't listening or had gone to sleep.

"Yes," said Howard, "jacks were and still are drinking fools, but some lumbermen have been known to squander their millions as easily as jacks spent their winter stakes. One lumberman who made a fortune hewing down white pine left his grounds near Waters, Michigan, inclosed by a wall of liquor bottles cemented together—bottles he and his friends had emptied." (Howard pointed this wall out to me on a trip to Lower Michigan, but we were not impressed by it. We prefer rail fences.) "Another lumberman, Robert Dollar, worked his way via lumberjacking and logging and lumbering into a fortune. The Dollar Steamship Line was his final investment. Dollarville, Michigan, is a town that bears his name.

"Many towns in Michigan were named for lumbermen, pioneers, or mining men. Deranson, a landing not far from Wellsburg from which Dad shipped many carloads of logs, was named for Dad. The landing was a nameless place at first—just a siding—but, as the number of log shipments increased, they found it necessary for convenience in making out the bills of shipment, to have it named. The spelling of Anderson was rearranged. The resulting name for the siding was Deranson.

"And Lewiston, Michigan, a community which developed after Uncle Jenson started a sawmill there, was named in his honor. That was back in 1892—boom white-pine-logging days. Wealthy lumbermen and sawdust towns were thicker than hair on a dog's back." Howard began unlacing his boots.

"Sure rich lumbermen was thick, and do ya know why they was rich? I've heard it said of Michigan lumbermen that they stole into the pine timber; they stole the timber; and they stole out'a the pine timber. I heard, too, that government pine was cut without ever being paid for. There was so much of it done the loggers called it 'moonlight loggin.' The cuttin' and haulin' was done at night. Sometimes daylight come along too soon and logs had to be left in the woods till the men could work another night. In between times the logs could be stolen by other moonlight loggers, because the loggers in possession of the pine was the owners. Bet that was excitin' loggin', stealin' one another's stuff that rightful was the government's."

That story, which Ed swore was true, now reminded Baldy

of tales his father told about pioneer farming in Lower Michigan. "The farmers came with only one objective," Baldy said, "and that was to clear the land for raisin' crops. They did more damage to forests than lumbermen ever thought of doin'. These pioneers didn't have the time or didn't want to log the big hardwoods and much of the pine. They used what logs they needed for their buildings and found other means of gettin' rid of the trees that kept them from plantin' crops. As quick and easy as they could, they felled the trees crosswise one another; and, usin' the smaller trees that burned fast, they built fires under the big trees.

"It took weeks to burn the big green trees. If they had'a logged their land it would'a taken much longer, though. When only the black logs was left, the neighbors brought their oxen to the future field. Hitchin' the oxen to the logs, the farmers rolled the black logs into piles along the edge of the cleared lands. The pioneers called these methods 'loggin' bees.'"

"I've heard that the smarter logger-farmers did their land clearing by windrowing," Howard said.

"Windrowing. What's that?" I asked.

"Windrowing was a glorified mass felling of thousands of feet of timber by actually felling only one tree. Long rows of trees would be notched in such a manner that they would fall at an angle toward the center of an area selected by the logger. Then, at one end of the line of trees to come down, one was selected that would strike the next tree in line when it was felled. As the end tree came thundering down on one end of the row of notched trees, it carried the one in front of it crashing down, too. So each tree felled the one ahead. They say the noise was deafening. For a distance of a quarter of a mile one tree felled another until the last tree notched came plunging down, all shattered to splinters and worthless for logs.

"Weeks later, when the trees had dried enough and the wind was right, a fire was started. The whole section—now a series of rows of felled trees—became a blazing inferno. When the big blaze burned out, there was the land, devoid of the hindering forest, waiting to take a few handfuls of seed in patches between the charred and blackened stumps." Howard sighed. "A remarkable procedure–windrowing–and loggers are called timber thieves, even today."

The jacks yawned. Ed looked at his watch. "Eight o'clock boys. Guess I'll hit the hay."

"Yeh, us, too. 'Daylight in the swamp' comes early," they said, and went back to the bunkhouse, their boots squeaking on the snow. Tex slid inside when they opened the door, and he then curled up in the warmth of the stove.

Ed shook down the ashes, added more chunks to hold the fire all night, and pulled the office door tightly shut.

I uncurled from my cramped position and closed our curtains. I heard Ed wind his watch, drop his boots to the floor, and climb into his bunk. Tex's nose appeared under the hem of the curtain. When he was not sent back to the office, he flattened himself and crawled under our bed.

Howard hung up his socks above the stove. He blew out the lamp. In the last glow of light I saw a wisp of purple smoke curl out of the blackened lamp chimney. A faint whiff of kerosene smell drifted to me. A long patch of silver moonglow shone on the rough plank floor.

Ed snored in deep sleep. After a few minutes the mice came scampering out to play along the shelves and on the counter. I heard them squeak. Howard sighed and tucked the covers around our shoulders. I fell asleep.

Booze, Broads and Brawls

All loggers and most lumberjacks are excellent storytellers if one can persuade them to let loose with their seemingly fantastic tales. As soon as I became a seasoned logger's wife in the eyes of the real old-time jacks, they quit hoodwinking me. When they came in the office after supper for snoose or candy, or lounged on the bench in front of the bunkhouse in nice weather, they told me what they knew and had heard about Michigan's fabulous lumber industry. Their stories, profanely punctuated and colorful with Swedish, Finnish, or other dialects, were absorbing to listen to. But those tales would almost never get beyond the censor's desk. The following tale is a composite, chronological, censored, and verified version of all the stories told by the lumberjacks who came and went from our camps.

"Tim-ber-r-r!" in our Tula Camp was the fast diminishing echo of the "Tim-ber-r-r" cries which had cut the cold morning air in the white pine forests of early logging days. The cadence of this warning cry had risen in volume ever since the first pine of Michigan rolled from its stump in the winter of 1835. The crescendo of lumberjacks' husky voices reached its fortissimo in 1873.

"Tim-ber-r-r" sounded the major chord from which all the melodies of the sawdust towns were composed. The crack of falling pines, the ring of the anvil, the thundering of logs shooting down the fast-flowing rivers, and the lumberjacks' songs blended with the whine of the mill saw, the creak of the carriage, the blast of the mill whistle, and the voices of the millworkers to compose the greatest of all logging and lumbering songs of the Middle West. It swelled with the times. Then gradually the song diminished in volume.

In the early 1830's great tracts of timber in Michigan were bought from the government by lumbermen migrating from the once heavily wooded states on the East Coast. Short-sighted officials who wished to have their fair state progress and become quickly settled gave away, practically, the emerald jewel in Michigan's crown of natural resources, because they imagined the land would be more valuable when it was cleared for farming. As a result, lumbermen rushed into Michigan and bought valuable and extensive stands of white pine timber for almost nothing. Seeing their timbered areas in the East fast reducing in size, the lumbermen looked westward and saw, rising from Michigan's rich ground, forests of pine beyond their most elaborate expectations. Straighter, taller, and more abundant white pine than they had ever seen or imagined stood before them in Michigan.

Gust, a bent-over, hard-working sawyer who could sharpen a crosscut saw better than any jack we knew, once told me, "I heared tell how the timber hogs got so thick in Michigan they couldn't see the pine. Some of them timber barons just kept goin'—'crosst the 'Sissippi River, 'crosst the prairie—till they ended up way out West."

First to Michigan came the early cruisers who traveled in canoes along the rivers from the bay regions to locate the stands of timber in relation to the earlier government survey markings. The cruisers' importance has never been fully understood or appreciated, except perhaps by the lumberjacks who sang the praises of these early cruisers to me in superlative sonorous cuss words. (If the lumberjacks have heroes, which I doubt, these early cruisers would qualify for the honor.) Yet, if the cruisers had not been willing and able to endure hardships of the same nature that the first explorers experienced, the lumbermen could not have heard the glowing descriptions of white pine, larger than two men's arms, with hands clasped, could encircle. They would never have learned of fast-flowing waterways, ideal for the necessary transportation of pine sticks and beginning at the very heart of the forest. Neither would the timber barons have heard tales extolling the immensity and denseness of this wilderness. The lumbermen believed the timber could not be cut in a century, it extended over such a vast area.

With fabulous tales to guide them and cruisers' estimates to fortify the fabulous tales, the lumbermen began the invasion of the Michigan wilderness in the 1840's. In ever-increasing numbers they came from the East Coast by waterways of the St. Lawrence River, by Lake Ontario, and by Lake Erie to Saginaw Bay in Lake Huron. From there the lumbermen penetrated into the heart of the forest, and floated logs down the rivers to sawmills located strategically at the mouths of the fast-flowing rivers. Lumbering, as Michigan lumbermen knew it in 1874, was without precedent.

The hordes of jacks who first invaded Michigan's wilderness with axes were colorful, robust men who are practically gone today—a group of men whom hardships could not discourage. They thrived best on beans and salt pork. They felled trees from daybreak until dusk in bitter cold, and slept on boughs from dusk until daylight. This sort of life they would endure for months without respite.

Confronted with a forest to be logged, the jacks of olden days, according to Elof's story, "gulped down a pint each of booze in two swallows, spent a night at Madam's Parlor, got 'pocked up' by gettin' tromped in a street brawl, and then, with a saw partner, headed up the river in a canoe for the timber. Or they followed a blazed trail to a log camp."

Elof, in an ill-fitting jacket with ragged cuffs, bought a square of plug tobacco and warmed himself, front and back, at our office stove. "Missus, them jacks slashed pine like trees was butter," Elof said, "singin' and cursin' for a whole winter and stoppin' only when they got hungry or couldn't see no more for the dark. Like this they worked and chopped 'til spring. Then, when snow started to melt, rollways high as mountains along the river was ready to be flipped loose into the churnin', swollen river. White ice chunks and logs floated down the river in the foam."

When the shanty boy sang out, "Roll out, daylight in the swamp," these lumberjacks climbed from their three-tier, double pole-bunks into which they had crawled clad only in the red flannel underdrawers. Around an open fire, the smoke of which drifted in the general direction of a hole cut in the roof—the only exit for the acid-smelling stuff—they ate their fill of beans and salt pork and washed them down with gulps of strong tea.

Lumberjacks who lived in this manner, as Elof said, "ain't no more." Death by falling timbers—the widow-makers—came quickly in the forests. The white, swirling waters of broken log jams choked out the life in many hearty birlers. Excessive drinking and street brawls claimed a good share of the cursing, wild lumberjacks of early logging days. Those who saw the sun shine full on Michigan's hills and valleys, those who saw the logs roll and churn down its rivers, and those who saw its towns carpeted with ankle-deep sawdust drank lustily to their good fortune.

" 'Tim-ber-r-r,' cuttin' da blue-cold mornin' air vas a varnin', 'bevare da falling timber.' 'Tim-ber-r-r,' yarring da bar glasses, meant drinks for efrybody in da saloon. A live, hot-blooded lumberyack vill pay da bill. Yah, a yack damn glad tu be alive, kickin' and drinkin' tu his own goot luck!" Oscar (a big Swede with disordered gray hair, a loud voice, and piercing blue eyes) vowed, as he pounded his fist on our counter in the office and made the cards jump.

The life was rough for those who went to the woods, but

they quickly learned all the dangers. "Da empty boots of da fresh dead tella dem often, and dey donna forgit it quick," Emilio, another old-timer said.

But lumberjacking was their choice. Many of the jacks who had logged the timber from the East Coast states came with the lumbermen to Michigan. But these were not enough men to shear the crop of timber from Michigan's land. Jacks came from Canada—French-Canadians who wore bright red sashes, red stocking caps, and red underwear, chopped-off pants held up by wide suspenders, and heavy calked boots. Chicago, teeming with men of many nationalities, supplied the camps with other woodsmen, immigrants from the Slavic and southern European countries.

The smooth talk of lumbermen's agents about the dense stands of towering white pine lured many immigrants of all nationalities in search of personal freedom and wealth into lumberjacking. Eric, a tall, blond Scandinavian who wore a perpetually sad expression, told me this story one late afternoon while he cut the paper around his tin of snoose with a cracked and rough fingernail.

"Vy did I come tu da voods? Vell, a big Svede come tu our shurch von vinter night. A bad vinter it vas. Ve shivered in little clothes. Ve sucked our tongues vile ve listen tu da big Svede tell of land of plenty, land of much vork, yiant trees in great forests, vast fields—to be had for asking und vorking. Ve listen, even da young vones. Ve talk all night. Not next year—or next—but next year many of us come here. Ve come tu cut da trees—not for much like ve be told, but for nuttin'. Ve no go home. No money. Ve stay in da forest."

Thus, directly to Michigan, a land covered with white pine, rich in natural resources, and certainly the fairest of that day, there came a host of Danish sawyers, Swede jacks, and Finn woodcutters. By their prodigious hewing and by their willingness to endure severe hardships and prolonged isolation they made fortunes not for themselves but for the lumbermen.

The tremendous lumbering industry grew with every new

invention. Crosscut saws for felling trees replaced the razor-sharp axes. Felling trees, laboriously accomplished with rhythmical strokes of an ax, progressed at a much more rapid rate. Before the hot ponds were kept free of ice all winter by steam pipes running from the mill, the jacks who felled the trees in winter came to work in the sawmills during the summer. When they could work in the mill winter and summer, the mill men left the cutting of the timber to the robust, howling lumberjacks.

With the invention of the endless bull chain and the improvement of the band saw, logs started through the mill with greatly increased speed. The sawdust flew faster and faster. The speed with which the crosscuts ripped Michigan's pine from its lands, and the bull chains fed pine logs to the greedy mills, was astounding. And it was close to vandalism.

Immigrants who were lured to the woods filled their packsacks with warm clothes. From Chicago's wharves they boarded lumber schooners that made their way between barges and tugs into the open waters of Lake Michigan. Expectant men watched the civilized horizon of mill stacks spouting black smoke, of leaning lumber piles, and of tall schooner masts, smooth into the even outline of the wilderness of Michigan—vast stands of ancient pine.

In the mill town of their choice, at a logging company office beyond the tumult along the docks, the lumberjacks inquired about directions to the timber. They acquired a sawing partner and hoisted heavy packsacks to their shoulders. Along newly cut forest trails they set out for camps. At night they slept in warm hollows cushioned with spruce boughs. They got up with the sunrise to continue their treks to logging camps and the timber to be cut. They followed blazed trails around fallen trees and over rivers, on through the deep wilderness. These trails they followed would become, before long, well tramped down by the passing of many lumberjacks in heavy boots and by mule skinners toting in supplies and chuck for hundreds of jacks.

Eric told me, as he slid his cud of tobacco from one cheek to the other and spat the chocolate-colored juice at the ash pan under the office stove, "Deer vould sniff da scent of us und run avay down da trail. Porcupines scuttled up da trees. Black bears vould lope off, crashing da dead branches of da fallen trees. Und at night da timber volves vould circle round us und howl a time or tu. Ven dey see da flames of our pine-knot fire, dey vould sneak off into da black, black forest."

At the end of the trail stood the camp buildings, flooded with sunlight and alive with activity. According to Battista (the short, dark-haired Italian with rosy cheeks and leather-brown skin who drove our frisky gray team), the walking boss, with a cruiser's help, located a source of sufficient water for the crew, laid out the camp site, and sent advance crews in to build the camps, in order that the buildings would be ready when the jacks began to arrive. "Oxen browsed in da back by da barn. Already yet me smella da peegs in da pen. Da blacksmiths, dey hammer avay to makka da beeg wheels. From da forest da jacks, dey cutta da poles to makka da bunkkas. Some, dey file da saws and sharp up da axes. Me? I cutta da cordwood for to keepa da bones warm when da wind howl and da snow she filla da chinkkas by me bunkka."

About the time that Battista took to the timber, the bean-hole in the bunkhouse had been done away with and meals were being served in a separate building connected to the bunkhouse by a passageway. Barrel stoves of enormous size gave the jacks more warmth than open fires. "No more da longga shelf by da wall for sleeping in," Battista told me, while he bought a suit of warm underwear from our supply in the office. "Bunkkas was double with three layers likka you say, 'three beds ahigh and two beds along'? Da bunkkas, dey was push up close togeder. We gotta climba in at da end. We no likka da muzzle-loader bunkkas. We no likka da double bunkkas, eeder. We cutta da snortin' poles and putta dem under da blankets, too, to keepa da partner on his side of da bunkka. Da Finns, dey stay by demselves, no wanna sleepa with me—me no Finn."

True enough, the jacks were clannish, especially the Finnish jacks. Besides being clannish, many of the Finns were sullen and silent. At Tula Camp they seldom loosened up enough to talk to me any more than to ask for their wants with one or two words. They merely grunted for "shirt, blue," "snuff," "big pants," or "fire time"—meaning that they quit.

Tauno, who spoke with an inimitable Finnish brogue, was an exception to these general characteristics of most Finnish people. He talked much and dramatically. On Sundays, without fail, he wore an archaic vest that was embroidered with an elaborate design, now faded and soiled. The edges of his threadbare vest were frayed and mended in places, but this big Finn, Tauno, wore his old-world relic with pride and dangled a heavy gold watch chain from its pocket. In the evenings when Tauno talked with us, he leaned against the counter and rubbed the short pink stubs of two fingers on his sleeve. The look of indolence on his broad handsome face was, in reality, extreme fatigue from his strong days' work.

One Sunday evening Tauno fumbled with his watch chain, which gleamed in the pale lamplight, and slowly, in husky words, told us how crews of jacks had worked along the rivers to be used for the drives, cleaning the river bottoms and removing snags, boulders, and fallen trees that might cause log jams when the pine sticks would come floating down the rapid-flowing rivers in the spring.

These men did their work thoroughly, Tauno said, because they knew that an easy drive depended on a clear channel and a river bed free of obstacles. While the river bottoms were being cleaned, sawyers cut tote roads and cleared skidding trails. Mule skinners brought in wagonloads of supplies. Many more jacks arrived, and the bunkhouses filled with sawyers, blacksmiths, teamsters, swampers, and loaders.

But not until the last bunk in camp was occupied did the jacks turn out to the timber. Then the sawyers swung their limber, glistening saws and keen-edged axes over their shoulders, and the teamsters bawled to their oxen. Into the tall pine forest they strode—a colony of men moving into centuries-old timberland to remove the trees and roll them into rivers where they would be carried on to mills and used in a thousand different ways.

Year after year tall trees crashed down through the forests ringing with the sounds of hard-swung axes. Sawyers limbed the fallen pines and sawed them into logs. Plodding oxen patiently skidded logs and pulled sleighs piled high with pine out of the forest to the "gees" and "haws" of drivers in languages unintelligible to anyone but the oxen. Swampers cut "watermarks" in the bark of these logs to aid mill hands later when the floating logs were sorted at the mill ponds. The marked logs were piled in decks on skids near the rivers' banks. Every day the rollways grew higher and higher. Thus the jacks Tauno told about worked from daybreak till dusk in snow knee-deep.

In camp, blacksmiths worked constantly to keep sleighs and rigging in repair and to fashion the oddly shaped shoes with which they shod the oxen. The cook and cookee got up long before daybreak to mix hot cakes, fry salt pork and potatoes, and brew the coffee. The bull cook, too, was up before daylight to build fires, light lamps, and fill the depleted piles of wood by all the stoves. One of the first to leave the warmth of the bunkhouse in the early morning was the barn boss.

Every man in camp made full use of the wan winter daylight during many long winter months. Those who did not were not welcome, but those who did were dead tired at dusk. The "bucking board" told why the jacks crawled into their bunks early. And it also told of the continual rivalry between the many sawing gangs and skidding crews. There, on the bucking board, each day's logs—those sawed, skidded, and decked—were marked down by a wise foreman who encouraged the rivalry. The log count grew faster with competition between gangs. This competition also provided subject matter for bunkhouse talk. Concerning the rivalry between saw gangs of different nationalities, I remember the story Gust told me as he peered over my shoulder at some of my sketches one stormy afternoon when the jacks laid in.

"There was sure a lota fights between the Swede gangs and Finns. Swedes and Finns both was known for their ability as woodsmen. And they was always fightin' to see who could cut the most logs, skid the most logs, and deck the most logs. The foreman always kept them fightin', because that way he got more logs cut; and that was good. Of course there was a lot of accidents. Some men got killed. When a jack was hurt, the hat was passed among the jacks to pay any doctor bills. Once, Gust recalled, 'after a Finn got killed by a widow-maker, the hat was passed to pay for the funeral. 'Und who bane killed nu?' asked a big, husky Swede.

" 'Toivo Laaksoharju had a widow-maker catch him. It broke his neck. The jacks are giving fifty cents to bury him,' the foreman explained.

" 'A gude idea!' said the Swede. 'But vy yust fifty sant? I give yu von dollar und yu go bury *tu* of them Finns!' "

As the winter wore on, according to Tauno, the Finn who wore the embroidered vest every Sunday, the lumberjacks in the log bunkhouses began to grumble and quarrel, count the logs they had cut, and figure their stakes. Some talked of their lady friends, and others drooled at the mention of liquor. Tauno claimed some got so dry by spring their skins began to crack. With March or April came the boom of breaking ice. The jacks then quickly forgot their fighting and put aside their grievances.

The rush of rolling, jamming logs, sliding down the banks announced that breakup was at hand. The logs plunged into the rivers, shot upward amid a geyser of white spray, and settled on the turbulent rivers and then floated down the waterways. Some logs which had been piled on the frozen river went downstream with the ice. But the foreman would usually wait until exactly the right time for the big drive when the river was deep from the spring thaws. On skids, as the log decks were piled, once the foreman flipped the key log loose from the snow- and ice-bound pile, the whole deck would come pounding down with tremendous power and thundering noise. Logs—pine sticks—churned and swirled down the rivers to mill ponds and eventually would be dragged into the mills by the bull chains.

When the big log drives had begun, husky birlers, "log cuffers" Tauno called them, grabbed their peaveys and followed the logs down the river, sighting logs caught on river bends which, if not removed, would cause serious log jams. Sometimes logs would wedge so tightly in an hour and pile so high by the force of oncoming logs that the use of dynamite charges was the only possible means of breaking the jam. To avert these bad jams, Tauno said the jacks strung flybooms across the river farther up to retard the progress of the logs. These flybooms, logs linked together by chains, saved many hours of perilous work freeing a log jam when the key log causing the jam could not be found quickly.

Eric, the sad-faced Swede, told me, while he waited for Howard to get a new handle for his ax, "Ven da ice vent ut und da drive started, ve yacks vould follow down da river, too. Ve sleep in tents or under da stars round von big camp fire. Ve ate von big meal from da shuck vagon in da mornin', und anoder ven ve come to da camp at night. At noon ve vould boil a can of tea to varm us. All day ve vatch da logs go down da river. At night ve come back to da camp site, vet, cold, und hungry like volves. But ve vas glad it vas spring, und ve vould sing around da fire, und den go to sleep quick ven da volves begin to howl."

All the water in the rivers of Michigan could not extinguish the lumberjacks' elation at being on their way to the mill towns. These log cuffers sang with gusto the songs that immortalized lumberjacks and their doings, which I am told all rivermen sang. "Log cuffers rode the foam when the logs thinned out."—But sure-footed, daring, and nimble as the birlers were, many lost their lives between the grinding logs and in the treacherous swirling waters. Most jacks considered losing their peaveys a far greater disgrace than losing their lives. But Eric said, "Da log cuffer, da vearer of da calked boots, vas mighty damn glad to hit da sawdust towns *vit* or *vitout* his peavey."

If logs were not properly stamped before they left the rollways, and sometimes when they *were* marked, there was always a chance that they would never reach the mills. Some logs became stranded; others were stolen. Thieves devised means of diverting logs from the river channels and marketing them as their own product. The most difficult work of driving logs down the rivers, the old-time jacks Gust and Elof said, was done by the men who followed the drive and rolled stranded logs into the water. Millions of white pine sticks reached the sorting ponds long after the main drive. Millions more logs were lost in the river beds and never recovered. Thousands of lumberjacks in hundreds of camps sweated and toiled to send millions of feet of pine logs yearly down Michigan's rivers to towns in whose harbors schooners rocked, waiting for navigation to open so that they could pull anchor and be off to exchange their cargo of pine boards for more immigrant lumberjacks.

After breakup, the jacks who did not follow the log drives down the rivers to the mills tore loose from camp and, half-running, made their way to the sawmill towns to join the birlers. There they squandered their winter's earnings in true lumberjack style. Hordes of jacks swarmed into the towns, Tauno said. They jammed into the saloons and the brothels. Howling and singing and cursing, they terrified the townspeople with their bloody brawls. There were scandalous goings on when the lumberjacks from the North hit town.

A chip on the shoulder of a jack was a dare for another jack to flip it off. The one who dared to flip off the chip would cause a fight on the spot. The bloodiest brawls were always for no reason at all except that jacks would always fight at the "flip of chip"—the suggestion that one jack was a better fighter than the other. The honor of being the better man, regardless of the issue, went to the jack who could stand up alone the longest. No holds were barred and anything was legitimate fighting. Calked boots disfigured men's faces for life. Tips of ears were bitten off. Fingers were broken. Flesh was torn. And when a jack went down, he could easily be killed, because a brawl never ended, claimed Tauno, as long as one jack felt like fighting. Jacks became known for their ability to fight as well as for their expertness with axes and saws and for their agility on rolling logs.

I remember well a short, stocky Austrian sawyer, Bastian, and his Italian partner, Emilio. The Austrian would tell an experience and Emilio would confirm it by, "Dats sure ting, Missus. You count on dat." Or the fabrications of Bastian the Austrian were discounted by Emilio. "Don't you believe dat, Missus, dat jist hear-tell." But some of this Austrian's inferences have been confirmed by more men than Emilio. It is generally known that jacks let broads in bawdy houses quietly relieve them of their hard-earned stakes. Talented ladies, of questionable virtue and accomplishments just as questionable, entertained the jacks with money to spend. Lonely jacks, susceptible to seduction after long cold months in isolated logging camps, paid any price for the company of women.

But there is this story told to me by Eric as he hung his crosscut in the rack on payday at the beginning of breakup at Tula Camp. "Some of us vanted to take our sveethearts for buggy rides down lovers' lanes in da nice summer nights. So ve'd rent a rig from da livery stables. But twern't no use. Efery time ve give da horse free rein, da danged ole nag ud head for da bawdy house. Efery time da same ting. Dem city slicks ud been drivin' dem nags all vinter vile ve vas cuttin' down de trees."

No group of men have consumed in their time so much booze as have lumberjacks. An entire winter's earnings has been squandered over one bar. "What a jack did not spend when he was conscious, someone took from him when he sprawled dead-drunk under a table," Ed said, "just for the rent of the six-by-four-foot space he covered."

Jacks have paid dearly for their drunken sprees, but the price meant little to them when they had tasted nothing stronger than tea all winter. "A mighty thirst can rise up in a man," Ed, our foreman, said, "when he sits in the wilderness, dead tired, on the hard deacon seat at the foot of his bunk. He recalls with relish the terrific wallop he got once from drinkin' a bottle of whisky. He remembers the unloosenin' of all his joints, and his light-headedness on his last spree. In camp, miles from a saloon, a guy's memory of the taste of whisky is always best." Jacks are known today, as of yesterday, for their insatiable thirsts—booze, broads, and brawls.

During spring breakup long ago, there were no infractions of laws committed by lumberjacks. Jacks were respected as men by whose prodigious labors sawmill cities existed. Officials generally overlooked their transgressions—rightly called "brawls." Some towns even provided elaborate entertainment for the jacks to welcome them and the influx of their currency. This practice only started bloody brawls, I was told, and reminded the townspeople that the jacks had arrived from the north. In the mill towns the jacks were welcome until their money disappeared; then they were tolerated. Some jacks found their way back to the camps where summer logging was possible. Some rode the rods to the harvest fields out West. Others got along the best way they could until the winter logging season began again.

One dry, hot autumn day at Tula Camp, I sat outside in the shade of the cookshanty talking to Eric. The chips and bark lying thick in the slashings by Arvid's shack had curled in the continuous heat of burning winds and sun. The twigs, leaves, and branches of felled trees crumbled, dry as tinder. The bark had dropped from a big log behind the office, and from the stumps their moisture steamed. The water had evaporated from the "low spot," leaving the muck under the planks cracked and hard-baked. The usual cool of autumn had left our forest. The jacks worked for Howard in the oppressive heat with sweat rolling down their chins and soaking their shirts. The horses came into the barn splotched with sweat. For many nights the men had found it impossible to sleep in the heat of the bunkhouse. They found the night air cooler outside the confining tarpapered walls. Eric, the old jack with the sad face, had on this day been forced to leave the woods because of the heat. In the sweltering afternoon heat he drank a dipper of cold water and splashed some on his disordered gray hair and hot face.

"Yust right veder for forest fire.—Like time I vork by Manistee. Ve smell da smoke. Mule skinners, dey bring news of fire in slashin's nearby. Ve yacks culd do nothin'. Ve yust vatch da smoke roll up und vish for da vind to shange. I see da deer sniff da air, raise a vite tail und leap off to go by da river. A black bear, he crash in da same vay to da river. Da vild animals is smart, dey know da vay da fire come. Dey know der is danger. Ve yacks yust gotta *sit* und *vait* for da vind to shange vith ashes und smoke floatin' round us."

During the logging days that Eric remembered, no one showed much concern about forest fires. With such vast stands of timber—an inexhaustible forest the lumbermen thought—the few acres (years afterwards estimated at well over two million acres) which were in the path of the flames were allowed to burn without attempts to check the devastation being wrought. Thus strips of black fire-scarred trees cut the virgin stands of green pine and hardwoods in an amazing jigsaw pattern.

Forest fires raged in those years until the winds ceased fanning them, and the flames died away. Fires burned, smoldered a while, and then were beaten out by rains. Or the forest fires burned until they warmed the waters of an inland lake. Some fires burned until a ridge of sandy, rocky, barren hills gave forth nothing more to burn.

Other fires that worried Eric and many lumberjacks long

ago burned steadily and approached sawmill cities located on the edge of Lake Michigan. There they consumed lumber piles, sawmills, wharves, and homes. Booming lumbering cities became devastated by forest fires which burned rapidly, fanned by sudden high winds. But the sawmill towns were rapidly rebuilt. The mills resumed sawing pine logs into boards. Logging continued with its former gusto. To lumberjacks like Eric was left the work of salvaging what timber remained in the charred pine trees. Of all logging operations, felling a burned-over forest was and still is, Howard says, the most dangerous. Sometimes the job is even too dangerous to be ever attempted, and the trees are never salvaged.

Every year since the 1840's, lumberjacks—Finns, Italians, Austrians, Frenchmen, Swedes, and Danes—cut the timber from the edge of Michigan's pine forest. The lumbermen pushed inward and northward year after year, letting a wealth of golden sunshine light the black forests. Lumbermen paved the way for the railroads which gangs of men laid while the cries of "Tim-ber-r-r" still hung on the air.

Simultaneous with the cutting of millions of feet of white pine came the growth of many lumbering centers and the making of million-dollar fortunes. These boom times continued for many years. But the fabulous estimates predicting that Michigan's pine would last over a century were shattered in the late 1800's. "Tim-ber-r-r" in the tall pines became diminuendo in a minor chord which was heard in the Upper Peninsula region while the big lumber companies stripped the last of Michigan's big white pine forests from Lake Superior's shores. Then lumbermen like Howard's relatives began logging the abundant hardwoods.

Today "Tim-ber-r-r" cries still echo in the maple and birch stands around Tula and north of Wakefield. They compose the finale of the unrecorded symphony of logging in Michigan.

And thus ends my composite tale of all the logging tales I heard at Tula Camp.

Breakup, Square-hewed Pine

Winter is long in the Upper Peninsula; but just as suddenly as winter had come to Tula Camp, the snow disappeared. Raw March winds blew. Crows filled the air with their raucous cawing. The air became warmer. Gentle rains fell. Daylight time lasted until after supper. Where high drifts of snow had been piled, there remained only dirty, honeycombed patches, reminders of long cold winter nights and sparkling sunny days, howling winds and coyotes, cracking fires and snow-laden branches, blinding storms and radiant moonlight—things of the past, but sure to be again. This was the beginning of "breakup," known to loggers and lumberjacks as those weeks during which the last patches of snow melt on the protected sides of rocky knolls, and muddy logging roads dry enough to be passable.

When a black cloud of crows settled in the scraggly pine behind the barn, our jacks began preparations for leaving camp during breakup.

"That flock of crows is a sure sign breakup ain't far off," Ed said.

Charlie on the tractor had for several weeks pulled trucks up slippery hills and out of soft places along the woods roads. The highways had become difficult to travel, and frost heaves made load-hauling restrictions necessary. A few light loads could be pulled out of the woods, but Howard decided to close camp and wait until warm rains and winds had cleared away the snow, and sunshine had dried the muddy roads.

Emil and Ole and other sawyers finished cutting their last strips. Steve skidded out logs that had been lost in deep snow since January. Most of the men seemed anxious to hang up their saws and axes—Silver Street beckoned. The blacksmith Paul let the red coals in his forge turn black and dead. Teams puffed and snorted, lacking their early fall and winter pep, as they pulled the jammers in to Paul's shop to be repaired later in the spring. Florida, the cook, made good use of odds and ends of supplies and didn't order fresh meat that week. Cowboy scrubbed the floors for the last time of snow and manure. Truck drivers came slowly up the slippery hills with their last loads of logs. Mr. Rogers stopped in the office on his final trip of the winter back to Mellen with the veneer truck.

"Good-bye, Mrs. Anderson. I won't see you and your lad 'til a good day from now. See thet ye take thet good care of yourselves. Been a fine winter for you? Now, thet it has. I can see for meself. Good day, now."

He shook hands with Howard and me, loosened his muffler, and climbed into the cab of the veneer truck. When I saw him again in October, he wore the same hand-knit muffler and hadn't changed a bit.

The next morning before Howard and Ed went to breakfast Howard flung the office door wide open and propped it with an ax handle. "Well, Julie, I guess this is the day we button 'er up 'til she dries."

All morning Howard and I made out checks and attached bonuses. The men brought in their tools, put their axes in the rack, and hung their saws in their places. Cowboy stacked unused bedding on Ed's bunk and stripped the jacks' bunks of their blankets to send them to the laundry. The jacks displayed much good humor. No regrets about buttoning 'er up were shown. Everyone began making preparations for leaving camp before dark. Christmas payday had been much like this, except for the change in the weather.

"It's been a good-paying winter, but I'm glad to have a

chance to relax for a while. Do you want to live out here this summer?" A sly grin crept over Howard's face.

"I should say not. I still have lumps on my arms from last summer's mosquito bites. It's been grand being out here this winter, but I'm anxious to see people, to get into a dress, wear lipstick, and gad about. I might even do a little serious housekeeping, like baking cakes and hanging out white washings."

"As for the men, as soon as they have squandered their money and the wet places are dry enough to work in, they'll be back at their old jobs. I'll have to drive out early every morning, and I won't be home until late. Will that suit you?" Howard asked.

"Sure. At least we *both* won't be eaten up by mosquitoes and no-see-ums. I can come out here for a day or an afternoon when the bugs are gone."

The jacks shaved off their whiskers and dressed up (all jacks do to dress up is put on tattered hats in place of their caps), buckled their packsacks around their worldly goods, picked up their checks, and found a ride to town. Most of the lumberjacks made no indication of when we would see them again. The Wisconsin farmers asked if we would be logging at Tula Camp next year and if we'd save a bunk or two for them. Ed and the loading crews and truck drivers understood that as soon as logging could be carried on again, they were to return.

Howard and I were the last to leave. In a short time Howard had loaded the few things we were taking with us in the car. Tex stayed behind with Axel, whom Howard had hired to take care of the horses and the camp while it stood empty.

When the last jack hit the road, we drove away. Melted snow water sloshed up with the weight of the car when we drove over the plank road. We skidded around on the slippery, muddy hills. A gentle rain splashed on the windshield, and the warmth coming from the car heater was welcome. When

Howard opened the car window to throw away his cigarette, damp air rushed in laden with the promise of spring—pale may-flowers, trilliums, spotted adder's tongues, queer-shaped Dutchmen's breeches, exotic lady's-slippers, leafing trees, balmy weather, song birds, and *arbutus!* Through the deep ruts of the field road, through Watermelon's gate, and out onto the bumpy highway we drove to Wakefield.

Breakup time meant to me the appearance of clusters of lovely starlike arbutus. Other spring flowers wait for warmer winds and sunshine, but not the hardy, exquisite arbutus. Snow is not all gone from our Michigan woods when the brave arbutus make their appearance.

I watched the sunny sides of the woodland knolls, near Brandt's Landing for weeks for the first delicate, sweet-scented arbutus stars. The pale-pink blooms hid from me under their shiny leaves, which made them harder to find than the shyest violets. The white arbutus were eager to greet the spring, and these I easily spied.

My bouquet of shell-pink and chalk-white arbutus had the most refreshing yet nostalgic scent I know. It reminded me of my father and his fishing creel, which upon his return from an early spring fishing trip could always be depended upon to hold, besides a rainbow trout or two, a bunch of arbutus tucked away in a corner and protected from trout smell by a quantity of its own waxy foliage. The starlike arbutus that bloom in the cedar-scented woods on knolls beneath the hemlock, pine, and leafless hardwood trees are my favorite wild spring flowers.

When their dark-green leaves had dried and twisted into crisp brown curlicues, breakup season was nearly ended.

Robins arrived and wild canaries and wrens. The air became warm, and gardens and lawns sprouted playthings left out by children all winter. Howard's trucks were all overhauled and the tractor tuned up. Every day seemed to him to be the day to start summer logging and hauling the logs from the mountainous decks along Tula Camp roads.

"The weather's sure nice. But it's no use to do anything until the swamplands dry so the sawyers can get to them, and the roads get hard," Howard reasoned. "Guess this is as good a time as any to see if Mr. Landon at Marathon Paper Mills will give me a contract for logs. Our contract with Bonifas has been filled."

That afternoon Howard came home with a contract it was to take us two summers and a winter to fill.

Brandt's Landing was so called because Mr. Brandt now owns the land on which the railroad company maintains the loading spur. The place was piled high with Howard's hemlock logs hauled from Tula Camp and decked there during the winter. It all dried before the woods roads did, and swampy places in the timber. This dry landing enabled Fred Brandt and his loading crew to begin shipping the paper-mill logs to Kimberly Clark before Howard was able to summer log. Knowing how much I enjoyed watching the loading crews at camp, Howard took me with him whenever he needed to be at the landing. There impressive decks of hemlock logs and drays of bark covered the grounds—thousands of feet of potential paper pulp and great quantities of potential tannic acid. Day after day the big jammer swung peeled logs up into the gondolas, pausing only while the men hammered in the stakes and wired the logs onto the cars.

On the landing in the early spring the crew worked continuously. When I had driven to the landing once during the winter, the trucks were slow coming from the woods. The men then had to stamp their feet and swing their arms to keep warm. Some walked to Fred's cookshanty nearby for a cup of hot coffee. Then they found time to tell tall logging tales, but not this spring. Nothing kept them waiting now, for mountains of logs stood at hand. Once a car was loaded, the men spotted another, and logs swung up to fill it. The never-ending process continued day after day, until the landing lay bare, with only chips and pieces of bark lying thick where the decks of logs had been piled.

Bark Peelers on Brandt's Landing

While the loading crew tore down the decks and Fred and Howard were busy, I watched the bark peelers work at one end of the landing, stripping huge pieces of bark from the logs and piling them all to dry in the sun. With an ax they cut the bark around the log at four-foot intervals along the log. Then they made a slit in the bark down the log's entire length. With a spud, they pried three four-foot pieces of bark away from the log and arranged them on the ground to dry in the sun. This drying process was necessary because green bark molds when it is packed tightly in boxcars—and molded bark is worthless for extracting tannic acid.

Later in the summer, when the bark had dried sufficiently to prevent it from spoiling, the bark peelers piled the big slabs on drays, which a single horse hauled to the boxcars. One man carried three or four curled slabs of dark-brown bark at a time up the slanting platform laid on the sill of the boxcar door. He packed the bark tightly along the side and back of the car until it solidly filled the boxcar. The bark loaders made many trips into the blackness of the car with bark slabs hoisted to their shoulders before they pulled the doors together and wired the car shut. And many carloads of bark were loaded and shipped to tanners in Milwaukee before the summer was gone.

To me it seemed that the procedure of loading bark must have been monotonous and dull, except for the brief rests when the red-faced men wiped the sweat from their faces and necks with big red handkerchiefs and lounged in the cool darkness on the shady side of the high log decks. Peeling bark is not as laborious as some woods work, although on those warm days the men peeling and loading bark probably wished to be cutting hardwood in the middle of the winter.

When I stayed home to bake and clean, I could see from our kitchen window loads of logs and carloads of bark being switched onto the main railroad line. Counting the log loads as they slowly creaked by, I could tell whether the jammer had broken down or whether all had gone as well as usual.

Some days I counted eighteen cars of hemlock logs, peeled and pale yellow in the sunlight, as they moved along the spur to be hooked onto the train of ore cars. On other days fewer loads were hauled, and occasionally only six or eight loads left the landing. Howard came home tired and discouraged on these days. But early the next morning he would eagerly return to the landing.

As I watched the never-ending activity of the loading crew and the bark peelers, Fred chatted with Howard and me. He told me that when the first hemlock was cut around Tula, the jacks peeled bark in the spring in the woods and piled it along the woods roads. Later, the bark was hauled on sleds to the landing and left to dry. Late in the summer it was shipped.

I have learned that now much of the tannic acid used for tanning leathers comes from South America in acid form. Still, each year some bark is peeled at Brandt's Landing. The men there are always busy, either unloading log trucks, making log decks, peeling bark, or loading bark and logs for shipment.

Instead of driving to where the men were working on the landing, as he could have done, one day Howard parked the car on a road parallel to the railroad tracks. "Fred's got the landing jammed with scattered logs. We won't try to park there today."

Howard went ahead of me to make a path through the tall grass and willow brush to the tracks and the landing. His long strides took him quickly far ahead of me. I followed at a leisurely pace the trail he made.

The day was bright with summer sun and warm. I responded to the weather and swung gaily along, not watching the ground. With big steps I strode over a slight knoll, never expecting in the whole world to see what confronted me. Suddenly I came upon a huge pine snake coiled in the sun on the flat surface of a rock directly in my path. The prominently marked snake was bigger than any I had ever seen.

This snake coiled in front of me was tremendous! Its black and brownish-yellow color made it seem more venomous than I knew pine snakes to be. Its size frightened me, too. As it writhed into a tighter coil and flicked its lightning-fast tongue I was, at first, fascinated; then I screamed, leaped directly over the mammoth writhing snake, and stumbled up the incline to the tracks.

Without looking back to see what became of the snake, I rushed to the landing, breathless and excited. By the gravel pit along the camp road, where dozens of brownish snakes came out to sun themselves, and on the sunny knolls at the edge of our timber, I had seen snakes slither out of sight in the leaves. But none of them were as big or as frightening as this hideous black-and-yellow snake.

At the landing, amused men greeted me. Some had their backs turned, but their shoulders shook visibly with uncontrolled laughter. Howard and Fred acted as if they knew nothing of what had happened. They were overly pleasant, and Howard asked solicitously, "What did you scream for?"

"You would too if you had to leap like I had to do just now over a black-and-yellow dragon that was spitting fire."

They laughed heartily. That they enjoyed my truly feminine reaction of screaming when I saw the snake coiled in my path made me furious. The joke was on me. Howard's keen eyes had spied the snake sunning itself far ahead of him. He had decided to let me discover it myself, he said. "I went around the snake and told the fellows on the landing what to expect. When you screamed, they knew why, and, of course, they got a good laugh on you."

Late in May, Howard came home one afternoon and announced with his old enthusiasm, "I drove out to Tula Camp today. The roads are all dry enough to start hauling again. Axel says the swamp's drying fast. We better find a cook. Would you drive over to Marenisco tomorrow for supplies? I'll send word to the truck drivers and they can round up the loading crew. I'd better find a blacksmith to repair jammers and get that new team shod."

Thus the planning for the summer logging began. The joyous welcome Tex gave me when I delivered the supplies to the new cook nearly made me change my mind and stay at camp for the summer. The memory, however, of the flies, heat, and mud at the Cedar Camp kept me from staying.

Axel and the new cook were joined by the old crew, except for the Wisconsin farmers and the ever-changing gangs of sawyers. Much to my surprise, Emil and Ole walked into camp one day when I was there.

"Good chuck, big timber, and old friends here, vy shouldn't ve come back?" explained Ole.

Activity at camp swung back to normal life in a summer logging camp. The cookshanty and bunkhouse doors and windows were screened. Flies droned in the sun while the horses switched their tails incessantly. The south wind brought the aroma of the manure pile to the men who sat on the benches in front of the bunkhouse during the cool evenings. The trees were now fully leafed, making the woods a symphony in shades of green, soothing colors lacking the stimulating effects of autumn's glorious array of lively hues and of winter's white sparkle. Trucks growled through camp loaded high with logs from the decks. When the decks disappeared, the drivers picked up fresh-cut logs at muddy skidways. In the heat of the summer the crew worked steadily but without the drive that the snappy cold winter weather produces.

All summer I went with Howard when I could, and he was glad to have the company. I tried to dress so that if he went to camp I could walk along muddy roads as well as through the woods and along dusty roads. By lapping the legs of my serge pants and tucking them into the tops of well-greased, waterproof boots, I had the perfect outfit for going through wet places or following Howard through the timber. My feet and legs were well protected against water, mud, stubble, and scratching branches.

With my pants legs flaring at the knees for easy walking, I am sure I must have provoked a polite smile from any who saw me attired in this strange manner. I will admit that my appearance was not comparable to that of the sleek, slack-garbed girls of today, their lips always bright with lipstick and their ears adorned with globs of gold, but I felt well dressed. Sometimes I wore my pants legs outside my boots. Then I probably looked less queer. During the winter I had worn snow pants, a warm shirt, and a pair of rubbers with leather tops, like those the lumberjacks wore, but white. To be warm, dry, and comfortable was one of my main concerns, winter and summer.

Colorful tales can be told about the products cut and shipped from the Upper Peninsula and about their uses. I had never dreamed that a Michigan forest product would be needed in its raw state on the other side of the world. But on one of our frequent trips to Brandt's Landing we came upon Jim Humphrey, a logger from Ewen, and his crew loading into gondolas the biggest squared white pine timbers I have ever seen or will see again. The men were loading the second car as we came to the landing. I wanted to drive home for the camera. Howard, for some reason, overruled that suggestion.

As we got out of the car Mr. Humphrey and Fred came over. "Big stuff you're loading there, Jim," said Howard. The timbers were so large that four of them filled a car. The hand-hewed pine easily measured five feet square on the big ends.

"Yes, this is the biggest white pine I've been able to find for several years. I looked all over for this and finally found five giants in a tract of pine timber near Lake Gogebic. It stood along the old Indian trail to Lake Superior. I don't think there's any more like it for size growing anywhere around here."

I watched a timber slowly ascend the skids. The jammer groaned. The rigging went taut—and held. Judging from the great size of the timber, I expected it to settle heavily in the car when it dropped. It fell lightly and easily into place. I was about to ask where the timbers were being shipped when I saw their destination painted on the side of the cars. Bold white stenciled letters proclaimed Cairo, Egypt, as the end of their journey! Cairo, Egypt! This did not seem possible, yet it was obviously true. These last big white pine which had been carefully felled, sawed into logs, and squared with sharp axes soon would leave Brandt's Landing on their way to Cairo, Egypt!

"What will these big timbers be used for in Cairo?" I asked Mr. Humphrey.

"Pine, lightweight and without knots, is perfect for ship patterns. That is how this shipment will be used. Forty years ago I sent similar shipments of white pine timbers to England and one carload of several timbers about five years ago, but these to Cairo will be the last. There isn't any pine of this giant size. Oh, perhaps a tree or two but no great number." I thought of the one standing at the edge of our timber. I wished I could have seen these pines before they had been felled. As the last timber slipped into its place on the third car, I regretted that I had not gone back after the camera. I have been sorry a hundred times since. Even though Howard was not impressed by these big square-hewed pines, I was.

The next day I heard a trainload of logs and ore cars creaking along the railroad spur. I looked out of my kitchen window and saw three gondolas loaded with square-hewed pine, an Upper Peninsula product, gleaming butter-yellow in the sunlight, slowly being hauled to the main railroad line along with cars filled with purple ore. These products from this land of timber, this land of iron, rolled along the tracks. The ore would go to Pittsburg, destined to become steel for bridges that span the world. The squared-pine timbers would go to Egypt, destined to become patterns for ships that take the "measure of all the seas."

Raspberries Grow in Slashings

The harried loggers expect to contend with many things. It is a not unnatural part of their jobs that fluctuating prices, equipment breakdowns, temperamental jacks, extravagent cooks, careless truck drivers, heat, cold, snow, mud, fires, and sudden death should all cause anxious and irritating days and nights. There are also fishermen, deer hunters, and bird hunters to be put up with. Loggers calmly endure all these troubles, but are most tolerant of the three last mentioned. And yet these sportmen types, in turn, show the least consideration toward loggers. And then, in the summer, in addition to the insect pests, loggers have to cope with berry pickers—the devout and insistent lovers of wild raspberry jelly.

Because Howard told me wistfully, with a boyish hungry look in his eye, that there was nothing to compare with the flavor of wild raspberry jelly, I decided to make some. Personally, even though wild raspberry jelly may taste better than tame raspberry jelly, I now prefer, at this late stage of the game, to make my jelly of tame berries. To explain why is simple, but to make other Range people agree is another matter which I shall never attempt. To the native son or daughter, picking wild berries for jelly is a wonderful Upper Peninsula summer sport for which they choose the hottest days of the season. They put on the queerest garb they can find—mechanics' coveralls, for instance, with generous patches of red flannel pinned to the seats. There is a reason, it must be confessed. While they are crawling on their hands and knees through the brush and over fallen trees and lost lagging piles, the glare of their posteriors will save them from being mistaken for deer by hungry poachers or shackers.

They roll up their sleeves and tuck the legs of their coveralls into the tops of long woolen socks worn in shapeless oxfords. Around their necks they tie large red kerchiefs to keep the bugs and worms from crawling inside their clothes, and they throw big scarfs over their new hairdos. To keep the sun out of their eyes they wear last-year's wide-brimmed picture hats. They anoint themselves with sundry mosquito repellents and reek to high heaven of the pungent-smelling oils. All this is just regulation dressing procedure for all berry-picking addicts. And I have seen even worse getups.

Onto Sam Brown belts they fasten pails into which they intend to put their berries. They take along plenty of containers for the berries they hope to find hanging in brilliant clusters from the bushes close to the roads. They take gallons of coffee and lemonade—the coffee for a bracer, the lemonade to soothe their parched throats. They fill picnic hampers with dozens of sandwiches and the rest of the chocolate cake they intended to have for dinner, because they are sure to be home in plenty of time with oceans of berries and can pop a fresh berry pie in the oven in a minute. . . . (They may prove to have plenty of time, but they are lucky if they can sit up without loud outcries over sore seats and lame backs. Some are lucky to get *home*—let alone make a pie, or even prepare dinner. I know, because I once went picking berries to make wild raspberry jelly.)

My experiences may not be comparable to those of other people. Anyone who ever goes berry picking where I did, however, will surely undergo a rather typical contact with rugged berry patches.

Someone told me the best berry spot in Gogebic County was north of Ironwood—down a certain logging road not far from Lake Superior. After I dutifully put on various pieces of queer clothing and packed myself a lunch, I loaded the car with lots of containers and started out. I followed the directions given me. To my surprise, I came to find that this logging road led to our old Cedar Camp! I swung around the familiar curves with confidence until I noticed that a car blocked the road which was obviously being traveled by log trucks—I had already met two. Now who but a berry picker would park a car *there?*

Knowing that the road shoulder was quite solid during the summer, I turned out and carefully drove around the car, wondering as I did so, whether the next log truck would have to head for the brush. But there was more of this nasty visitation ahead. On all the turnouts and parallel roads, cars were parked. I could see wide-brimmed hats bent over berry pails. No one waved a greeting. All were anxious to be the first to get to the biggest berries.

As I was going to pick berries on my old haunts, I decided to go where no one else would dare to pick. Up the steep, steep hill I climbed—beyond camp to the west near the old deer yard through the slashings, all that remained of a once dense forest. The muffler scraped jagged stones. Other noises I heard made me think Howard would have to replace the oil pan when I got home. At an old skidway I parked the car, smeared on mosquito dope, and began walking down the road. It curved through the cut-over land Howard had logged years before. (Raspberry bushes grow two years before they bear fruit, and even in two years the brush and weeds grow thick, too.)

Disguised among the berry bushes were prickly thistles and vines that made perfect snares for the unsuspecting. The sickening-sweet smell of flowering grasses, weeds, and vines and the indescribable odor of lush summer foliage replaced the pungent smell of conifers and the sharp clean smell of hardwoods. Giant mullein towered above the other weeds from beds of velvety gray-green leaves and displayed only several greenish-yellow blossoms on their tremendous spikes of potential blooms. Profusely growing Canadian thistles thrust out sharply barbed leaves. Some of the lavender thistle flowers had gone to seed. Fluffy thistledown floated away in the summer heat. Purple fireweed flowers with their exotic scent were twined around with the vines and pale trumpet-like blossoms of morning glories. Yellow touch-me-nots, whose seed pods popped at the least pressure to release ripe, black seeds, grew high above the tall grasses.

Beyond all this were berry bushes laden with their bright fruit that enticed me deeper into the slashings. The day was a scorcher. Every kind of insect in the woods was attracted by my brand of bug repellant. I untied the kerchief around my neck. That gave the bugs the chance for which they were waiting. They all crawled down my neck at once and began frisking around in my blouse.

The rapidly increasing summer heat of this eventful day sharpened my thirst and deadened my sense of smell. Red-brown thrushes fluttered in the thick brush. A yellow-speckled flicker hammered on a towering blackened rampike. A wild canary trilled from a scraggly hemlock tree, unruffled by my lumbering approach through the bushes, weeds, and brush piles.

Except for the bugs and heat, the first hour was enjoyable, and I picked several quarts of berries which I dumped in the containers at the car. As I dropped the juicy fruit into the pail at my waist I thought how refreshing it was to get away from town and tramp in the woods—or, rather, these slashings. Suddenly I tumbled into a hole up to my neck, my berry pail turned upside down, I landed on a prolific thistle, and my left ankle throbbed with pain. What I said couldn't, by any stretching of one's imagination, sound like, "Oh, dear me."

I crawled out of the hole, hobbled to another bush, and began again to strip bushes of their fruit. To fill the bottom of my pail took much longer this time. My ankle began to swell, and the thistles kept working through my coveralls to jab me viciously at frequent intervals. The berries seemed smaller. I tried to pick faster, but by picking faster, I grabbed more leaves and bruised the berries.

The sun became blistering hot. My hair felt plastered to my head with perspiration. The coveralls were hotter than a Dutch oven. What I needed was a cold drink and a sandwich. Back toward the road I crawled, over logs, around stumps, through the weeds and brush piles. But where was the road? Where I thought it should be, it definitely was not.

"Can I, who have tramped this area any number of times with Howard, be lost?" I asked myself, hoping to see a familiar stump or rampike to guide me back to the logging road. Nothing looked familiar. I turned and went in the opposite direction. That way, too, looked unfamiliar. I circled left and walked into a big patch of bushes with berries hanging heavily from the branches. Forgetting that I was lost, I rapidly filled my pail.

A crack of dry brush sounded!

I looked up! Directly before me appeared the face of a black bear! The look of satisfaction on his triangular face changed to a sneer, and then to a belligerent frown.

I screamed my most bloodcurdling scream! The bear nearly fainted, but recovered and plunged into the uncut timber.

"To heck with picking berries," I muttered, and stumbled in the opposite direction from the bear. I fell into a pile of brush, face down, and shattered one lens in my glasses. Quick tears of fright, anger, self-pity, and frustration wet my hot cheeks.

With clear vision in only one eye, I located the road and staggered to the car. A maddened wasp jabbed at my eyebrow as a parting friendly gesture from the slashings. Without touching the lemonade or sandwiches, I started home. Driving became increasingly more difficult as my left eye slowly closed. On both sides of the road other berry pickers' cars were parked. Turning and jolting over the rough road and swinging around the cars was none too good for my berries. A near-sighted glance into the pails revealed that with each bump my berries settled down and down until what had been five quarts became three. "Heaven knows what will be left when I get home," I thought in disgust. "Just worms and a sodden mess I suppose."

Interested berry pickers in my neighborhood met me in the driveway. "Did you find many berries?" they asked.

"Oh, about six quarts," I mumbled, as I squinted to see into my containers. My left eye was swollen shut, and I couldn't see much with the other one, lacking half of my glasses. These handicaps accounted for the extra quarts I claimed to have.

Later, when I talked to another berry picker, she asked with some skepticism, "Where in the world did you ever pick sixty quarts of berries?" I cannot blame her for such surprise or doubts.

Amazingly, my berry take, with numerous retellings, became six quarts, then sixteen, and finally sixty. Raspberries multiply faster than rabbits, it would seem—especially when lovers of the wild variety made into raspberry jelly are on the trail of a patch thick with berries easy to pick.

Entirely aside from my vicissitudes, that day I discovered that the best, sweetest, and biggest berries grow in the shade. There also—it must be added—lurk the biggest and most vicious mosquitoes, and bugs, and bears! And between the bushes there are nothing but thistles, touch-me-nots, nasty brush piles, twining vines, and wasps with irrational tempers and deadly accurate aim. These are what a berry picker can expect to encounter when he decides to follow the gleam and pick *wild* raspberries.

—But picking the berries is, of course, merely the first stage in making the jelly. The second step is to make the jelly—and make it jell.

While I soaked my sprained ankle in hot water, I mulled through the berries and found a hundred and eighty-four big ones, which I put aside for the pie I intended to make (but didn't—they were a mess of mold in the morning). As to worms among the berries, not being able to find every one because of my swollen eye and missing glasses lens, I reasoned that one worm was as bad as fifty, and I closed my eyes to them all.

I did manage to pick out the dry leaves and big bugs, though; and I mashed what berries I had—beyond the special ones reserved for pie. By tying and twisting the mashed berries in my whitest dish towel, I extracted the juice and squeezed the bag for the last drop. I had heard that positively no one would ever think of squeezing the berries for all the juice, because that thoroughness makes cloudy jelly. Perfect jelly should turn out to be transparent and jewel-like. Cloudy jelly tastes just as good as transparent jelly, but it is cloudy. Just what is jelly I wonder?—Something to look at or to smack one's lips over? Accordingly, I squeezed the berries to a certain careful limit and removed their precious juice. It was still the unorthodox but the quick way.

Next came the little matter of the Epsom salts test for pectin. From a cookbook written for beginners I read, "To one tablespoon of cooked juice, add one teaspoon of sugar and one-half tablespoon Epsom salts. Stir this mixture for several minutes. If it forms large flocculent particles, the juice will make satisfactory jelly."

"I wonder, what *are* flocculent particles?" The pain in my ankle kept me from hobbling to the dictionary to find out. "My mixture certainly looks queer enough to be flocculent particles."

Having made the test, I added the sugar to the juice according to the directions. I kept this mixture boiling until it jelled, and I consulted my cookbook again. "Pour the hot jelly into clean . . ."

Good heavens! I had forgotten to wash the odds and ends of glasses and fancy mustard jars I had saved all winter at Tula Camp. I flew around as fast as my protesting ankle allowed, gathered up all the jars I could find, washed them, scalded them, and filled them with the lukewarm jelly.

With the hot paraffin poured over each glass, I proudly surveyed my five filled glasses—with one partly filled. Sad to say, some dozens of empty ones remained.

I sighed; I bemoaned my messy kitchen and thought how nice my jelly was going to taste spread on hot buttered toast some cold winter morning. Howard would surely say that nothing could compare with the wonderful five-plus glasses of jelly I had made.

Then I began considering a number of other things; the chances that ninety-nine to a hundred the jelly would taste slightly like kerosene because I had heated the paraffin too hot, or that the jelly might be so thin we would have to drink it, and that my precious jelly might even be covered with velvety green mold when we removed the paraffin.

I considered my aching ankle, my broken glasses, my swollen eye, and my frayed disposition. I thought of how disappointed Howard would be when he came home from camp hungry, only to discover that there would be stale chocolate cake instead of fresh raspberry pie.

Dear, oh dear! All this day I had worked for just five and perhaps a half glasses of jelly. The kitchen smelled deliciously of the cooked berries, but what a mess it was! I should have looked to see what Webster had to say about flocculent particles. . . .

When Howard came home he jumped sky high at the mess I was in. His appetite for wild raspberry jelly must have faded then and there. But that wasn't all. When the bill came for repairing my glasses, said he, "Julie, don't you dare let me catch you ever again in a patch of raspberries. You go pick apples. My favorite jelly is apple. Haven't I ever told you?"

Our Trees Take Shapes

At Tula Camp all winter the small forest products had been piled intentionally along woods roads, where they could be easily reached in the summer months. I noticed when I went along with Howard on warm summer days to the woods that these piles grew smaller and smaller. Each day truckloads of lagging, ties, or mining timbers, careening and swaying out onto the highway, reduced the evidence of winter logging.

Laggin', the tall broad-shouldered Finnish fellow with the wild howl, who had swamped behind the tractor for Charlie all winter, now drove a truck hauling lagging to the mines. My brother Hayward, lured by the novelty of living in a logging camp, came to spend the summer working for Howard. Howard started him out swamping for Laggin'. A swamper is merely a helper. Working with a tractor driver, a swamper hooks the choker and chains around the logs and cuts the brush out of the tractor driver's way. He also makes skidding trails for the tractor. This precedure allows the tractor driver to speed up his work of dragging logs to the skidways. The swamper on a truck helps the driver load and unload his product. Jacks say that the man who helps a bartender is "swamping in a saloon," and they sometimes call a cookee the "cook's swamper." Ed, our foreman, referred to me as "Howard's swamper" when he thought I didn't hear him. The lumberjacks' term, "a logger's swamper," fits me perfectly. Being a logger's swamper is, however, certainly a far greater undertaking than being a truck driver's swamper in the manner the jacks mean.

Hayward, built like my dad, barrel-chested and tall, helped load many truckloads of lagging that summer, but I doubt if Laggin' considered him helpful when Hayward drove him out of the cab by the smell of the cigarettes Hayward rolled for himself. Laggin' told me after Hayward went back to school, "Those cigs your brother rolled was tea-leaf stuff from the cookshanty. I got so sick of their stink, I gave him one of mine. That real tobacco did the trick. He sure got sick. He was too young to be a-smokin' anyways. And just like I thought he would, he quit puffin' on tea-leaf cigarettes, too."

Laggin' and Hayward strained and sweated, as did other Tula Camp truck drivers and swampers, hoisting onto the truck the split pieces of six-foot cedar which were used to reinforce the underground passageways in the mines. They piled their truck high with these comparatively lightweight pieces. As they left the woods with their enormous loads, the lagging swayed and shifted and all but tipped off the trucks because the loads were so extremely high. I thought it was miraculous that these loads of lagging ever reached the mines at all. Strangely, a number of log loads were dumped during hauling season that summer; but I have yet to see a load of lagging tipped over.

Laggin', a tolerant fellow, gave Hayward a good workout that summer. Together they hauled lagging—and ties when the lagging piles were gone—and claimed the greatest number of loads hauled by any of the trucks handling short stuff. From the cedar swamps, the railroad ties, cut, peeled, and squared by Arvid the shacker, were hauled out to the road on horse-drawn drays by gypo gangs. From these piles along the woods road the truck drivers and their swampers loaded the ties onto trucks with platforms and hauled them to the railroad landings. Piled there along the tracks, ready to be loaded into boxcars, these hand-hewn cedar ties represented considerable skilled work. When the railroad needed them, boxcars were spotted along the tracks; and a tie bucker loaded the ties by hoisting them, one at a time, to his shoulder, walking up an angling platform, and piling them inside the boxcar. This process was slow and laborious even when the tie bucker jogged along at a reasonably fast walk.

"How can a tie bucker make a living working at such a seasonal job?" I asked when we had gone to the landing along the railroad at Ironwood to see that the tie bucker loaded the right ties.

"When a tie bucker isn't working up here he loads ties cut farther south. He's paid so much a tie according to the size of the tie—two or three cents for average-sized ties, and four cents for large ones.—He can easily make as much as eighteen dollars a day. Railroad companies usually pay the cost of loading their ties. They hire a tie bucker who can travel from landing to landing to do nothing but load ties. Sometimes loggers agree to load their own ties, as was the case when I bucked ties for Dad and landed in the hospital with a crushed nose and a cracked skull. That was a bad break for both Dad and me."

In that short summer season of 1940, thousands of Howard's ties disappeared into the black boxcar doorways and were later delivered at treating plants to be dipped in creosote. Later, laid on cinder beds under strips of steel, these cedar bolts from Tula Camp helped circle the world with efficient transportation.

Loads of hemlock logs for paper pulp went to Brandt's Landing during the summer, too. Maple and birch logs were hauled from Tula camp to Marathon Mills on the outskirts of Ironwood as well. These logs went to help fill our new contract. The logs felled and skidded in summer were not as clean as those cut in the wintertime that were covered with white snow. Skidded through low, wet places, the logs cut that summer collected much mud, and it dried on them as they lay piled on the skidways.

For getting logs out of the mires on the edge of the swampland, the teamsters used skidding pans shaped like dirt scrapers. They pulled backwards. The front ends of the logs (which would in the ordinary process, straight out, have dug into the muck and made skidding a difficult process) now rode on the skidding pan and slid easily over the mud, the mire, and spongy leaf mold. The use of skidding pans was to Howard that summer what the use of big wheels was to white pine loggers who also logged during the summer months —an ingenious method which enabled loggers to log the year around regardless of the mud.

Howard occasionally received orders from the mines for trestle legs, fifty-foot spruce poles that were not easily found. He received one such order that summer of 1940. For several days Howard searched our timber for enough sixty- or seventy-foot spruce trees to make several loads of trestle legs. He carefully blazed the trees and sent a reliable gang of sawyers in to fell them. Taking care to fell these tall, slender spruce trees so that they would not shatter or split, the sawyers had the trestle legs ready to skid in a week. Over an especially cleared skidding trail that had no sharp curves Charlie dragged out the long straight poles with the tractor and left them on the widest skidway. Emil and Ole stripped the twisted bark from them. Then the loading crew rolled the poles onto the trailers which Howard had to uncouple and adjust for the additional length necessary to haul these extraordinary long pieces of timber. Watching the crew load these pitchy and sharp, pungent-smelling poles broke the monotony of the long hot summer days for me.

To insure the safe arrival of the trestle legs at the mines without causing other drivers trouble, Howard called the State Police and notified them in advance of the hazard these long poles would be. When the trucks hauling their unusual loads reached the highway, State Police, trim in appearance and efficiently alert for circumstances which might cause danger to car drivers, escorted our loads of poles to the mines. These poles, carefully unloaded to insure against their splitting, were used to support ore-car tracks suspended over the storage ore dumps and thus served the wide economy of our great country.

The summer days went by quickly and pleasantly. In this part of the Upper Peninsula, some folks say (a bit facetiously), "Summer is the Fourth of July in these parts." I have seen a "spitting of snow" in July myself, so I know they are not far wrong. We doubly enjoyed the summer season because we knew it would be short. We wasted none of the hours of sunshine. Howard customarily left Wakefield at five-thirty in the morning and worked all day with his men in the woods, with the exception of the time he spent at Brandt's Landing or at Marathon Mills in Ironwood. I did my housework in the early morning to be free to go with Howard in the afternoons, or whole days, if I chose.

One Saturday afternoon late in the summer, when the jacks had all left to cool off and clean up in Hurley, the foreman Ed, Howard, and I put a load of pine logs on a truck to clean up a skidway. Howard did the loading while Ed and I hooked. This load of pine has a special significance for Howard and me, because today we have in our Cape Cod home pine wainscoting made of the very logs which we loaded on that hot summer afternoon. The pine logs were sawed into lumber at Marathon Mills, dried, and piled safely in a garage to wait until our dreams of having a home of our own would come true. Not until eight years later were we to have the joy of seeing this white pine adorn our own four walls.

As usual in the summer when there is no deep snow to contend with, the crew spent one afternoon moving Arvid's shack. John Steffin, another shacker working for Howard, had selected a new spot in dense cedar and had been moved the week before. Arvid, the crippled shacker, had stripped the timber of all its short stuff within easy walking distance for him. When Howard found a new location for Arvid, sawyers quickly cleared a site for his shack. Warm sunlight flooded this spot in a thick stand of cedar, as the trees crashed down.

Charlie dragged the trees away and bulldozed the stumps to one side. Then he disappeared down the woods road and returned with Arvid's shack on skids. Slowly he crawled closer and closer to the new location and finally hauled the shack off the road into the clearing. With little difficulty, the men removed the skids, and Arvid's shack settled down on the soft black earth as if it had been built there. A narrow creek, offering clear sparkling water, bent itself around until it trickled by not over fifty feet from Arvid's doorway.

A truck bringing Arvid's grindstone, his sled, and other cumbersome things like the wash bench and tubs, brought Arvid, too. He lumbered up to his new front yard and looked around. "By yee, I got gud spot nu. Vater near by. Road near by tu. Und ders plenty stuff tu cut fer year or tu. By yee, I got von svell place here."

A clump of cedars growing close together made it a secluded spot at the curve of the road. This piece of timber bordered the section line. Beyond, not far to the south, I could see the black crowns of tall pines above the tops of the hardwoods. Towering above all stood the big pine. The smell of the cedars—the makings of ties—mingled strongly with the smell of birchbark torn from the trunks of trees rolled out of the clearing. The odor of fresh-turned leaf mold, rotten logs, moss, and dried leaves was heavy in the air. The men called to one another above the drone of Charlie's tractor. Arvid found his shovel in a pile of tools and began throwing the sod against the shack in preparation for the winter.

"Gonna be von long vinter, Missus. Rabbits got ticker fur den last year. Von got caught in da brush pile. I let him loose. I see he got much tick hair. Dis vinter be cold und long." Arvid proved to be a truthful predictor of the season to come.

Several weeks later, yellow leaves began fluttering prematurely to the ground. The grasses turned brown, and I missed the robins' songs. One cool evening while we lounged on the porch of our apartment in Wakefield, a lumberjack came looking for Howard.

"I heard you got good tie makin's down by Tula way, Mr. Anderson. I cut short stuff. C'uld ya set me up in a shack? Like to be by meself. I cut lots of stuff for ya cheap. Make fine ties. Peel posts for ya. Split laggin' good, too. Stay as long as ya want me ta," he pleaded.

"Where did you work before, Jack?

"Come from Minnesota."

"I'll see what I can do for you. Find a place to spend the night, and I'll take you out to camp in the morning to look the stuff."

"Yah. Thank ya, Mr. Anderson. I be here five in morning and go camp."

The jack liked the timber, and Howard had his men build him a shack. Several days later Howard asked me to fill this new shacker's grocery order. I bought his supplies and took them to Howard in the woods.

"Come on," he said, "I'll drive out to his shack and dump these things right now while I have time." I moved over, and Howard slid under the wheel. We bounced down the road to the last turn out. Ahead was the shack.

Howard carried a big box of supplies down the fresh-cut path to the shacker. I followed along with an armful of small things. Without greeting us, the shacker held the door open for us. About his place there still hung the smell of new wood and tar paper. A fire crackled in his tiny cookstove. His coffee-pot gurgled and puffed out the strong smell of shackers' brew.

"Got everything ya need?" Howard asked, while he pushed the only bench over to me so I could sit down. I looked around, comparing his shack to Arvid's.

"Guess ya need a couple more stools," observed Howard.

Glowering at me from under shaggy gray eyebrows, he blurted out gruffly, "I got 'nuff stools. Jis got too gol-dang many weak-kneed wimmin' clutterin' up my shack!"

I got out of his shack in a hurry and waited for Howard in the car.

"It's a good thing you left, Julie," Howard said when he

came back to the car. "That jack sure hates 'wimmin.' 'Won't put up with none of them thar skirts messin' up any shack of mine,' he said. Guess he'll calm down once we get out of here. He seems like a good worker. Quite a hustler. He's got a lot of stuff cut already."

The loading crew began decking logs for early winter hauling. The lagging piles disappeared entirely. All the ties were hauled. The empty places beside the woods roads told a story of continued summer logging activity. The mosquitoes and tiny black flies left with the first light frost. Being in the woods with Howard during the day was now much more enjoyable for me than in the heat of midsummer. The early evenings were cool and refreshing to the hot, sweaty lumberjacks. Many yellow and brown leaves covered the sun-baked ground. Milkweed pods dried, split, and curled, letting out silken parachutes that carried ebony-black seeds away on faint breezes.

Silver-gray fireweed spirals adorned the roadside. Crickets, hidden in the drying grass and weeds, chirped their monotonous autumn song. Goldenrod turned tobacco brown. Mullein spikes, exhausted of their prolific blossoms, stood sticklike at awkward angles. Clusters of burgundy-red sumac seeds clung to scraggly dwarf trees and glowed with sophisticated warmth against bright blue skies.

The early fall days slipped by quickly. Vibrant colors flamed in the usually sedate green woods. Chill winds swirled and tossed the falling leaves. Leaf at a time, like impish strip-teasers, the hardwood trees shed their bright leaves until they stood stark-naked—black tracery against the steel-blue sky. Sharp whining winds blew through the bare black trees, bending and whipping branches, snapping twigs, and piling leaves in heaps against mossy logs and in rocky gullies. In contrast to the indiscreet hardwoods, the evergreens remained warmly clad in luxurious green velvet needles.

Late in October, Howard and I moved back out to Tula Camp. I spent several mornings washing dusty and flyspecked windows in the office and in our adjoining bedroom. I replaced the cracked mirror above the washbasin with a new and bigger one. On our bed I smoothed a bright new spread. At the two small windows in our room I hung new crisp curtains. In the top drawer of our tiny chest I stowed away several new blouses and a number of pairs of warm socks. On top of the chest I spread a clean scarf and laid out my comb and a box of face powder. I checked my left-over sketching supplies; found a book I hadn't read; and dusted the alarm clock, wound it, and set it. Thus began the second winter at Tula Camp for me.

Fall rains fell—cold, icy drops that hung along the twigs and tree branches and sparkled in the fleeting sunshine that occasionally broke through the dark skies. Gossamer spider webs, which were invisible in the tall grass before the rain, became spectacles of prismatic jewels hung on platinum filigrees. Cold rains washed summer dust from roadside trees and tore away the last dead leaves that clung to bushes and branches. These chilling, drenching rains kept the jacks in the bunkhouse and made more work for the bull cook who scrubbed the floors. I sat by the stove and read or did some drawing. Howard made out van sheets for the new men and got the books in order.

"I wish the weather would turn cold and freeze up pretty soon. The men are gettin' restless. I suppose a number of jacks will end up in Hurley if the rain doesn't quit." Howard, too, was anxious to see the sun.

The weather cleared and turned colder. The men went out to the woods again. Jacks began the trek back to Tula Camp, looking for a good cutting chance, good chuck, and high prices for sawing. Tex, after faithfully following me about for days, began greeting old lumberjack friends and making new ones. The long bunkhouse gradually filled with men again; Steve the teamster, Ole and Emil the sawyers, Laggin', who howled, Clarence, Ray, Joe, Cowboy, and others who had worked for Howard the winter before and during the summer returned to their old duties.

The cook, a new one whose name was Swede, set more places at the two long tables. Cowboy the bull cook unpacked dozens of clean blankets just back from the laundry and made up all the bunks. He piled dry wood in all the racks and along the walls by the stoves. Fires, crackling in the barrel stoves, radiated cheering warmth. Appetites grew to be enormous, and chuck disappeared rapidly from the bowls and plates. Each morning heavier frost told of colder nights and the onset of winter.

The cold days, sometimes bright with sunshine, but more often dark under cloudy gray skies, were ideal for logging. Cries of *Tim-ber-r-r!* filled the air. As I walked along the woods roads, flurries of snow buntings flew up from the frozen brown earth to the nearest low branches of the hardwood saplings. With the dark gray of their wings and heads blending perfectly with the leafless birch and maple branches, their white breasts made them look like huge snowflakes being tossed by a rude gust of wind. The green moss and pale lichens showed brightly against the blended browns and grays of the tree trunks. The sharp smell of cedar and hemlock and the acrid odor of wood smoke replaced the summer scents of lush-growing foliage, wild flowers, grasses and weeds.

All the jacks and Howard, too, made the most of the gradually decreasing daylight time and came in from the woods tired and hungry.

Log prices went up and up—especially veneer. These months—the latter part of 1940—were the real beginning of World War II. Penokee Veneer had gradually begun to produce more and more materials used in making planes. The bulk of their veneer was exported to England for use in assembling mosquito bombers and manufacturing British-made propellers. These aircraft veneers—made from logs cut and hauled from Tula Camp—were exported under Lend-Lease. They were specially priced, which made the profit we realized on our veneer logs greater than ever before. Because we were paid more for our logs, Howard invested in another tractor and two new log trucks and trailers.

Besides the veneer, the jacks cut pulpwood hemlock logs which were hauled to Brandt's Landing as before, and loads of Number Two hardwood went to Marathon Mills to apply on the contract Howard had made during the summer. The shackers worked steadily and piled up cord upon cord of lagging, and peeled and hewed hundreds of ties. Log trucks rolled by the office with regularity. It was no unusual sight to see four trucks parked outside the cookshanty at one time—each driver having a cup of hot coffee before starting on his haul to Mellen, Ironwood, or Brandt's Landing with his heavily loaded trailer.

The crispness became freezing cold. Frozen mud made the roads full of ruts where the tractor blade did not smooth them down. Wisconsin farmers returned to Tula Camp to saw for Howard. A flock of ducks stayed one night on the pond beside the camp road. The next day they continued their flight southward. I watched their V-formation waver raggedly across the sullen gray sky. Above the fringe of hardwood tops they became fainter and fainter. Behind the dark crown of a tall pine they vanished. That afternoon a dismal rain turned into fine flurries of snow that covered the dead leaves and stuck to the rough bark of the birch trees. By suppertime the snow had thinly covered most of the ground and roads in the open places. We lighted our lamps early that night and banked the fire with solid chunks.

"Spittin' up a dang good storm," Ed said as he wound his watch. "Guess I'll put that extra blanket on me bunk tonight."

"Where's that heavy underwear of mine, Julie? Maybe I ought to put it on tomorrow." Howard tossed an extra pair of socks over the line above the stove to be warm and ready for morning. Through the south window of our room I watched snow stars drift down and pile up on the sill outside, while I climbed into warm flannel pajamas. Another winter logging season had begun.

Part Four

"Sound Shots"

Deer were always plentiful around Tula Camp. But during all my tramping along the skidding trails with Howard down by the cedar swamp and at the far edge of the section, I had never shot one. The deer only flipped their white tails in my face. By the time I had gotten in position to shoot, they were safe in the next forty. Howard was anxious that I should have the thrill of shooting a deer. He thought it would be great sport to have me come in from the woods to camp able to chalk up a kill.

One Sunday he called several lumberjacks who wanted to hunt into the office. A light November snow covered the ground. The jacks and Howard talked over what they thought was the best area to hunt in. They organized a drive to begin at the edge of the field and cover the territory back of camp. This drive was to be made to me to make certain I bagged a deer.

We checked our guns and left the office in high spirits.

"All ya gotta do, Missus, is keep yer eyes peeled. We'll scare up a big fat buck for ya. Ya can't miss. Nothin' to it. Jist watch for a big rack a horns. Then aim ta kill him dead."

The men went on ahead. Howard found a low stump for me and checked my gun once again. "Now, Julie, all you have to do is keep your eyes peeled. There's sure to be deer passing within good shooting range. This is your chance of a lifetime to get a buck." He left to join the jacks on the drive.

I sat quietly and patiently for a long time. My seat got cold through my flannel shirttail and snow pants. First I sat on one hand and then on the other. All the time I kept my eyes peeled (so I thought) and waited for *my* buck to show his horns. . . . I waited—and waited. I sat on my other hand a while and waited minutes that seemed closer to hours. Finally, I became so cold that I slid down off the stump, laid my gun on the stump, and jumped up and down to warm myself.

I heard a sharp crack in the brush behind me! I grabbed my gun and looked for *him!* Out of the corner of my eye I saw him for a fleeting moment. He leaped high over a brush pile. He was gone! My thrill of a lifetime had escaped me.

In very low spirits I waited a while longer. Then I stiffly made my way back to the plank road. There I found fresh deer tracks not more than one hundred and fifty feet from the stump upon which I had waited. I followed the tracks along the road and saw where the deer had bounded off into the slashings just behind my stump. I was ashamed to admit that I had let being cold interfere with keeping my eyes peeled. Dejectedly I returned to camp.

Back in the office, Pete asked me, "What happened, Missus? Did ya git struck with buck fever?"

"Yes, Pete. I guess you'd call it that." I let the matter drop, thinking buck fever would cover my predicament nicely. But Howard came along and exposed my failure as a huntress.

"According to the tracks I found, the deer saw you first and circled around you. You didn't keep your eyes peeled. He couldn't smell you. The drive was planned to eliminate that possibility."

Then I admitted I had gotten cold. "I just slid down from that stump for a minute to get warm, Howard. It would be my luck to have *my* buck come along at that time. Which would you rather have, a frozen wife and a buck with a big rack of horns, or no buck and me?"

"You, of course." The big hug and kiss he gave me warmed

me and forgave me. But it seems to me that keeping one's eyes peeled and freezing to death is the only way one can ever see and shoot a deer before he escapes into the timber.

I have never hunted deer since then, not because I missed a good shot, but because, with the invasion of our logging areas by the down-state hunters, I feel much safer at home. This feeling is contrary to all the statistics which claim that most accidents occur under one's own roof. Yet even the memory of the smell of choice venison steak smothered in onions can no longer lure *me* out to the woods to be a target.

As I left the office to go to the cookshanty for breakfast the morning after I missed my buck, I saw Steve leave the barn with Paddy and King. I could not believe what I saw. Howard's horses were always well taken care of. The teamsters kept their coats glossy and their tails and manes brushed. But never before had our horses been dolled up in red ribbons and bows. Both Paddy and King sported braided manes and tails with generous pieces of bright red cotton woven in and out. Perky red bows adorned all of the horses' rigging.

"Where's the parade, Steve?" I called out, as I walked down the road to get a better look at his team. At that moment from behind the barn Mulligan drove the gray team gaily decked out in the same unusual manner. Both teams looked like prize horses on their way to be judged at the state fair. Molly, a white horse the gypos used to skid mining timber, came out of the barn last. She was prettiest of all the horses. The ribbons in her white mane and tail were vivid crimson splotches. In this gay regalia, she reminded me of a ring horse from a circus. The teamsters had dressed themselves as gaily as they had decked their horses. Every jack wore his brightest plaid shirt, and pieces of red flannel covered their dark caps. Howard, too, I noticed, had selected a new red mackinaw to wear to the woods this morning.

"Beginnin' now there'll be a parade of deer hunters 'round these parts," Steve called. "We sure ain't takin' any chances on gettin' plugged full of holes, Missus."

"Sure enough, Steve, I know it's deer-hunting season. But why did you have to tie Molly up in bows, too? Surely the hunters wouldn't shoot at a *white* horse."

"You sure don't know these down-state hunters, Missus. They think loggin' horses are fine fat deer, the white ones bein' albinos. One hunter took a shot at a bus full of people last year. If them city guys even hear a twig snap, they take what they call 'sound shots.' Jist can't trust 'em. They don't believe in these here No-huntin'-or-trespassin' signs—or they can't read. They drive right past them signs, 'n park their cars on woods roads in the way of crews movin' jammers. And danged if they don't leave their cars in gear and locked for good measure. A guy has to rig his self up like a Silver Street gal—or get shot plum full a holes. It's just about got to the point where loggers ought ta close their camps, lock the horses in the barns, and get outta the woods durin' huntin' season. You better stick close to camp today, Missus. I'm tellin' ya these hunters is wild."

The teamsters and their gaily decked horses disappeared from sight as they rounded the curve beyond the barn. I decided to take Steve's advice and stay close to camp. I lingered in the cookshanty after breakfast and drank coffee. I watched the cook make half a dozen pies in no time. He put a roast in the oven for dinner. Then he dumped a pile of potatoes into a kettle and began peeling them. Howard came in for a cup of coffee and let in a draft of cold frosty air.

"Did those three hunters I let sleep in the bunkhouse come in here for breakfast this morning?" he asked.

"No, nobody came in. But I saw a couple of guys go out early right after I got up. I wondered who them guys were." The cook emptied the peelings in the garbage barrel by the back door. "Here comes a couple a hunters now. I thought you said there was three."

Two men in bright red garb pushed open the door. "Did you see our partner?" one asked. "We got separated from him about an hour ago. Think he went down in that swamp after a deer we got a glimpse of when we first started out."

"I went around that way, but I didn't see your partner. That swamp's a tricky place. Hope he doesn't get lost down in there. Does he know how to get around in timber?" Howard asked the hunters. "Better give these men some breakfast," he said to the cook.

"Sure, he's a good hunter. Never been lost. He'll be back all right. Just wondered if you'd seen him. We heard a couple of shots. Thought he might be on the trail of a buck."

Howard left for the woods again. The cook hustled up a huge breakfast for the hunters, which I noticed they consumed with apparent relish. I sipped my coffee and had another doughnut.

"Boy, I was more hungry than I thought," one hunter said to his partner as he helped himself to more bacon and another stack of hot cakes. When they finished they picked up their jackets and went to the bunkhouse.

I warmed my front and rear at the stove. Just as I was about to go back to the office an excited hunter burst in. "Say, where's—my partner—Joe? Did he—come—back to—camp?" he panted.

"Yes, I guess it was Joe who came in."

"Did Frank—come with him?"

"Two men—hunters—came in. They are strangers to me."

Overcome with excitement and fatigue, the hunter sank to the bench by the stove. "Gee, I guess I must have killed a deer then."

I gasped. "Good heavens, man. Don't you know what you shot at?"

"Well, *now* I do. Where's my partners? We'll go out and get the deer."

"No ya don't," yelled the cook. "You stay right here. Don't know what ya shot at, eh? Ya better stay right here with that gun."

The pickup slid to a screeching stop. Howard pulled the door open. At a glance he detected something wrong. "What's goin' on here? You all look scared. What's happened, Julie? You're white as a sheet!" I was speechless. The cook explained the situation.

"What do ya know. This guy shot at somethin'. Wasn't sure it was a deer till he come and asked around about his partners."

While the hunter mumbled an explanation, Howard suggested, "You boys better start out for home. You'll get killed by other hunters as careless as you seem to be."

Steve came in to the office that evening with French fire in his eyes. "See, Missus. What did I tell you. We sure will have ta get the heck outta the woods during huntin' season. The only thing that'll help would be a big snowstorm. That sure scares them guys home in a hurry. Won't be safe in the woods till them hunters git gone."

Steve's advice was good enough for me. I stayed in camp for the rest of the deer-hunting season. When the deer hunters returned to their homes that year, in their wake they left unclaimed doe carcasses, cases of empty whisky bottles, grumpy loggers, and angry townspeople. Tragically, some hunters were killed, some were wounded, and others saddened by the unexpected events which left them facing manslaughter charges. Their wild antics became the basis for tall hunting tales which the lumberjacks delightedly fabricated during the long winter evenings.

Suddenly the jacks at Tula Camp forgot the deer hunters. Talk got around about the big bird who was stealing caps from the jacks at another camp. According to the story Baldy told me, a huge bird swooped out of the sky several times one bright afternoon and soared out of sight, clutching a jack's cap in its talons each time. This daring thievery continued for a number of days. Superstitious jacks quit. Others had to go to town for new caps, and, of course, tell the story of how they had their caps stolen by this mysterious bird. Loggers made every effort to capture the thief, but none of the men assigned to shooting the bird had good enough aim to kill the robber.

Finally, one logger called the State Police. They stalked the bird for several days. The bird wisely did not appear on the days the police were hunting the thief. However, one sunny day this pirate bird swooped down from the sky to snatch another cap. An anxious trooper brought it down with a well-aimed shot. The stolen cap was returned to the jack to whom it belonged. Pinned to the nib gleamed a bright shiny union button. The bird, a hawk, apparently attracted by the gleaming buttons, stole no more lumberjack caps.

Union buttons were something new in the Upper Peninsula woods. In a few logging camps all the jacks were union members. But at Tula Camp no one sported buttons. Howard hired the sawyers and men he wanted, without discrimination for or against union members, with that typically American concept of equality which prompted him to hire men of all nationalities.

"What discussions go on in the bunkhouse about union activities, religion, or prejudices, are jacks' own affairs," Howard contended.

Only once did we ever wish we knew what the jacks thought and said. A carload of strange men visited the bunkhouse one evening. We thought nothing of the visit that night. When eight o'clock came, we blew out the light and went to bed. Later, when we heard the car drive away, Howard sleepily said, "I hope they didn't bring a bottle or I'll be short a man or two tomorrow."

They had not brought a bottle. In the morning at breakfast time Steve came in for a plug of tobacco.

"There's gonna be a vote taken about organizing your camp. See that notice by the bunkhouse door, Missus."

I looked out of the office window. Sure enough. A white printed card was tacked by the door. "Look Howard! Those fellows must have been union organizers."

Howard went down to the bunkhouse and read the notice. He came back slumped and glum. "Guess you're right, Steve. There'll be a vote taken in a week."

The following days were electrified by the shock that the union wanted Howard to operate his camp according to their regulations. The jacks, however, said nothing to Howard or me about the voting. Ed told us that they talked most about the attempts that organizers had made in 1937 to unionize all the lumberjacks and force loggers to operate under union contracts. From guarded tales Howard had told me before, I knew that most of Howard's present crew had worked for him or his father during that eventful summer. Now Howard refused stubbornly to talk about that strike. "We were all in it together. What you don't know, Julie, you can't broadcast." Instead, he dug out some old newspapers from the bottom of our davenport which revealed to me the reportable facts of a violent summer involving many more people than just jacks and loggers and organizers.

At last the day arrived for the vote to be taken at Tula Camp concerning a union contract. An unnatural tenseness showed on the faces of the entire crew. By suppertime some were downright grouchy. Tex tried in vain to coax the men to play with him. I felt uneasy. Howard never mentioned the voting. Glumly he sat, twirling a pencil and looking out the office window at the bunkhouse door.

Steve came in smelling like the barn. He took a new shoulder pad for King from under the counter. "Bet a ten spot those union guys don't show up."

No one had spirit enough to take his bet, but it would have been easy money for Steve if they had. Two hours slowly passed. We listened for a car to drive in to camp. I got nervous. Another hour passed. No one came.

The jacks left the office and went to bed. Lights in the bunkhouse disappeared. Ed dropped his boots extra hard.

"Guess we better go to bed, too, Julie." Howard blew out the lamp. We crawled under the covers and went to sleep.

The next morning I saw at once that the notice had been ripped off from beside the bunkhouse door. When Ed came back from breakfast I heard him tell Howard, "I guess those guys knew they couldn't swing a unionizin' vote here so they decided to skip the whole thing. I'm glad they did."

Never again was any attempt made to unionize Tula Camp.

Vigilantes

During the tense days while we waited for the organizers who never came, I turned the dusty, yellowed pages of the newspapers Howard had saved since 1937. From their headlines, long articles, and short local items I pieced together the events of that stormy violent summer the jacks and Howard refused to talk about.

I counted back the years and discovered that during the unpleasant upheaval in the Upper Peninsula that year, my sister and I were attending the summer session of college at Boulder, Colorado. She had studied journalism, while I had painted scenery for the little theater on the campus and pursued a course in interior decoration. We rode horseback along the mountain trails and enjoyed the beauty of Estes Park and the Black Hills on weekends. Our only slightly serious thoughts were about the disappearance of Amelia Earhart and the love of the Duke of Windsor for Mrs. Wallis Simpson.

Early in 1937, the general movement of workers across the land was toward the organization of collective bargaining groups. After spring breakup of that year, when Louie Anderson and other loggers were getting ready to haul the logs they had left for summer logging, over one hundred jacks walked out of Bonifas Camp Two, near Gogebic Station. Previously, in smaller camps in the county, a few jacks had quit after asking for higher wages. The Bonifas walkout was on a much larger and more alarming scale. Handbills were circulated in

Marenisco, Wakefield, Bessemer, and Ironwood, urging all jacks to let the Sawmill and Lumber Workers' Union represent them in collective bargaining for higher wages, single beds, shower baths, washing and drying rooms, recognition of their union, and free meals for all jacks looking for work. A small disagreement between a millworker at Marenisco and a log hauler about union membership began the strike which spread rapidly.

In a few days, five hundred lumberjacks had left the camps of Marathon Paper Mills, Bonifas Lumber Company, "Cedar Pole" Smith, McNicholas, and smaller operators like Fred Brandt and Louie Anderson. Every day more jacks left their jobs, but the organizers of the strike made no formal demands. They said they would not make any statement until all the jacks in the Upper Peninsula were out on strike.

With this ambition, the organizers proceeded to enter all logging camps and present their plans to the lumberjacks. After they took a strike vote, they herded the striking men into trucks which took them to the Palace Hall in Ironwood or the Town Hall in Marenisco. One group of organizers swaggered into a cookshanty at a logging camp and enjoyed a big breakfast without so much as offering to pay for it. Then, not at all graciously, they piled the men into trucks and left the logger with no lumberjacks to get his logs out of the woods.

At one camp the reason the jacks gave for striking was that they had not had meat for breakfast. The foreman told the boss of the complaint, who immediately consulted the cook. "Is it true these men didn't have meat for breakfast?" he asked.

"Yes, the supply truck was late," explained the cook.

"Well, what *did* you put on the table for breakfast?" demanded the boss.

"There were stewed prunes and apricots, fried eggs and potatoes, cooked cereal, pancakes and syrup, bread and butter, doughnuts, cake and pie."

"The jacks were probably hungry for excitement—and a good brawl—but certainly not for meat," said the boss.

Not many days later, over four hundred fist-shaking strikers sullenly marched from Bessemer to Ironwood in a demonstration which a spokesman said would be duplicated in a march on the state capitol if the strikers' relief demands for food, blankets, cooking utensils, and material for building shelters were not met. Strangely, many of the strikers were tough-looking men whom none of the operators in this region recognized as ever having been on their payrolls. Loggers have good memories and hire jacks who have worked for them year after year. But most of these men seemed to be strangers.

Log hauling came completely to a standstill. Loggers were forced to cease cutting logs entirely. It looked much as if there would be no bark peeling if the jacks did not go back into the woods to get the hemlock out in time. The strike spread to Ontonagon logging camps. Negotiations between the loggers and strikers were not progressing successfully. Activity remained stalemated in Gogebic County, and the organizers proceeded to Munising with their band of followers.

No further disturbances occurred in the Upper Peninsula except isolated acts of sabotage. The State Troopers then returned to their established barracks.

I asked Steve where he and other lumberjacks were during the early part of that summer, and he said, "Shucks, we let them roughnecks do the fightin'. Us lumberjacks were back in the woods, hidin', and loggin' for Hungry Louie and Cedar Pole. The organizers didn't take time to patrol the woods. They was busy pickin' fights. We said, 'To heck with that union business!' Louie and McNicholas, Cedar Pole, Fred Brandt, and all them other loggers always treated us O.K. We needed to eat, 'n we couldn't lay around all summer, so we kept on workin' far back in the timber. Let them bums do the fightin'. That's what we did, Missus."

The papers reported that by the middle of June drivers had begun hauling logs to the mills. The operators' civilian rights to use the highways without being molested were recognized. Howard began hauling logs for his dad, glad to get back to work even though he was taking quite a risk. The strikers were still eager to force their demands. A mob of them could easily overwhelm one truck driver if they wished to force him off the road.

In an attempt to prevent any such incidents, fourteen deputies assisted the sheriff's department and the State Police in patrolling the highways where trucks were hauling logs. The strikers made no efforts to stop the log haulers, although they damaged several cars carrying millworkers to Connorville. Hungry lumberjacks began drifting back down the camp roads and into the offices looking for employment. The real jacks wanted to go to work. But still no agreement could be reached making it possible for the operators to hire them without causing more trouble.

Loggers were willing to raise wages, provide single beds, and make the other improvements demanded of them; it was now merely a question of recognizing the so-called union. Their demands for free meals for jacks looking for work were "weasel words." Loggers were well acquainted with the two hundred or more camp inspectors that plagued their cooks and increased their logging costs. Free meals for honest jacks hitting camp and looking for a job had always been accepted as camp custom and never questioned.

All this idleness and hunger begot indiscriminate property damage, bloodshed in some instances, one death, (no one ever determined whether the man died of a blow on the head received from a man wielding a golf club or whether he died from a heart attack), violent hatreds, and then—inevitable retaliations.

Gradually truck drivers began to get the fast-spoiling logs out of the woods to the mills. There was no indication that the strikers would prevent them from using the highways. Most logging, however, was entirely crippled (except where a few men like Steve worked far back in the timber). Camps were desolate and empty. Idle horses, getting fat and frisky, stamped in the barns. The flour in the cookshanty became wormy and the lard rancid. Tall grass grew up along the logging roads and covered the paths around camps. It was late in June and the summer logging season would have been at its height, if there had been jacks to pull the saws through the trees and skid the logs.

Then one day two truck drivers pulled out onto the highway with loads of logs. Howard was driving the second truck. A mob of strikers were hidden in the woods along the road. They swarmed out of hiding and forced the first driver to unload his load and proceed with an empty trailer on to Ironwood. Howard saw the treatment the first driver received and decided to give his truck "the gun" to force his way through the men. He roared down the highway, he told me, but they stoned his cab, broke the windows, and bruised his shoulders and arms. They finally forced him to stop by jumping onto the truck, and while, one stoned him, another jammed on the trailer brakes.

Leo, "Shoopy," Isdebski, a deputy police, patrolling the road alone that day, saw the violent action and raced to the scene. In his hurry to disperse the strikers, he pulled the pin of a tear-gas bomb too soon. The bomb exploded before he could release it from his hand, and Howard and Shoopy were both caught in the fumes. Choking and gasping, their eyes smarting and blinded for a while by tears, they were abandoned by the strikers. Shoopy's hand was badly burned by the bomb. Bruised and sore—but still able to see—Howard drove him to Wakefield for treatment at the hospital.

They learned that other trucks had been unloaded on the federal highway west of Bessemer and logs scattered along the road. As a result, the next morning five strikers were sentenced to jail on various charges resulting from the stoning and dumping of logs. The newspaper made public the fact that only one man was a citizen, another had his first papers, while the other three were classed as "roamers."

During a fiery meeting that night the C.I.O. pledged its support to the Sawmill and Timber Workers' Union, as the lumberjack strikers chose to call themselves. At another meeting that same night, in an "Oust the Vagrants" movement, the Ironwood Chamber of Commerce called upon the government to take immediate steps to rid the city of vagrants and undesirables. The Rotary Club of Wakefield and the Ironwood Kiwanis passed resolutions with unanimous approval, deploring the un-American activities pursued in the woodsmen's strike and supporting the efforts made to maintain the security of personal and property rights. Copies of these resolutions were ordered sent to county and state authorities. Meanwhile, citizens watched additional State Police arrive at the police barracks.

Six weeks of smoldering fury flamed into a vicious battle at Marenisco the next day between hundreds of strikers and fourteen State Troopers and sixteen sheriff's deputies. Strikers stoned logging trucks and shattered truck windows and windshields as the trucks came through town. When the police came to the aid of the drivers, the strikers turned their barrage of rocks upon them. The troopers and deputies fought the strikers for over an hour to stop this brawl. The police emerged the victors, cut, bruised, and battered. They arrested forty-six strikers, brought them to the Wakefield post, and recorded their fingerprints. Investigation and their fingerprints revealed that all except seven of these men were aliens, some with criminal records.

At this point during my newspaper research, Howard came in the office. He leaned over my shoulder, planted a kiss behind my ear, and scanned the yellow pages with me.

"Ha, I see you're gettin' to the gore."

"Yes, and this last business of vigilantes chasing the strikers out of the country doesn't sound tame."

"Well, you can bet money on this, Julie. It certainly wasn't any Sunday-school picnic."

"Tell me what happened. I've already asked Ed about that

fight. All he said was, 'Ask me no questions, Missus, and I'll tell ya no lies.' Must be a touchy subject. No one wants to say anything about what happened."

"Well, I'll tell you some of what I know, but don't you ever quote me.—Dad and I got wind of the plans and left home without telling anyone where we were going. From others in the know we heard about a meeting at the Ironwood Fairgrounds. News of this meeting must have spread over the range like wildfire, for within a few hours vigilantes from Ironwood, Bessemer, Wakefield, and Marenisco had gathered at the Ironwood Fairgrounds."

As soon as the darkness of the warm summer night enveloped them in its disguising mantle, these determined sober men solemnly tied strips of white around each other's arms, armed themselves with ax handles, and proceeded to the strikers' headquarters in Ironwood.

While I soaked up the warmth of the barrel stove and pawed through old newspapers those cold winter afternoons, I learned that if loggers have no adverse weather or fluctuating markets and prices to contend with, on rare occasions they find themselves in the middle of a strike—or the threat of one. It was that way at Tula Camp. The beautiful setting for our logging activities provided only stimulating atmosphere for our continual struggle.

Sundays at Tula Camp

On Sunday mornings the cook delayed clanging the triangle until six-thirty. That delay should have given us all an extra hour of sleep. However, he could have dispensed with the ritual entirely, because every Sunday precisely at six Watermelon drove up the hill to collect the garbage for his pigs. The rattling of his truck and the banging of the garbage pails, accompanied by Watermelon's sonorous Italian arias buffa, awakened us. (This criminal disturbance of extra sleep on Sundays we suffered in exchange for the use of Watermelon's road from the highway across the field.)

With strains of basso buffo resounding in our ears, we took our time dressing for Sunday breakfast. It was sure to be something special like buckwheat cakes, pork sausage, and applesauce, in addition to the regular cereals, eggs, doughnuts, cookies, and all the rest of logging-camp breakfast fare.

After the sounds of Watermelon's industrious and songful activity had died away, the jacks appeared at the back of the bunkhouse to prepare for their Sunday "boiling-up." From the office window, Howard and I watched them while we waited for breakfast bell. Under the big black kettle that hung on a pole wired between two sturdy maples, the jacks kindled a roaring fire. They pumped pail after pail of water to fill the kettle. After breakfast, when the steam began to rise above the rim of the kettle, each jack waited for his turn to scoop out bucketsful of hot wash water into a tub on the bench nearby.

"Boiling up" on a Nice Sunday

Using a tin can punched full of holes at one end and nailed to a long stick, the jacks soused their underwear and socks around in the steaming sudsy water until they became the desired shade of gray. With quick twists the jacks wrung the steaming sudsy water out of their things and rinsed them in a tub of clear water. They squeezed out the rinse water and secured the clothes to a line strung between saplings, or they hung them on broken branches of trees.

There the clothes remained until they dried. Some hung limp from the line, and others twisted and clung to the trees. Frozen stiff in winter or flopping crazily in the spring breezes, these assorted shirts, holey socks, and tattered or patched underwear made a queer sight. By noontime on Sunday all who had washing to do, and the ambition to do it, had taken their turn at tending the fire, carrying water, and sousing their clothes in sudsy water.

Sunday afternoons at Tula Camp were usually quiet and uneventful. If Howard and I did not have urgent errands in Wakefield, we stayed at camp. The jacks slept, smoked, played cards, read, mended their mitts or socks or pants. I supplied them linen thread, big needles, and patches when they asked for such things. Other jacks greased their boots, filed their saws, or ground sharp edges on their axes. Some of the piece-makers worked, if the weather was exceptionally pleasant. A few jacks would return to camp late in the afternoon from their he-man sprees on Silver Street. They talked loudly, and staggered, and some collapsed completely upon reaching the bunkhouse.

One such Sunday afternoon, several jacks returned in unusually drunken condition. As I sat in the office on the high stool behind the counter adding scale reports, Mulligan, a red-haired, red-whiskered Irishman, cautiously picked his way up the hill to the office. He staggered, and his face showed the effects of heavy drinking. After several attempts, he pulled the heavy door open and, clutching the counter for support, he made his unsteady way opposite to where I sat.

"Say, Missus." He rolled his pale-gray and bloodshot eyes and jerked his head up. "Gimme a plug a tobacco."

I reached behind me, took a package of the kind I knew Mulligan chewed, and handed it to him. "I guess you have a hangover, Mulligan."

"No, no, Missus. Me no got hangover. Me jist gotta hang on."

To lumberjacks there are, evidently, degrees of drunkenness which I had never observed before.

North of Tula, on the east side of the Presque Isle River, stood a section of timber Howard wanted to cruise and perhaps buy. This section is a small part of the heavily timbered and vast wilderness region at the very foot of the Porcupine Mountains cut by crystal-clear streams and rugged with knobby hills and deep gullies. In midwinter several feet of sparkling snow covers the ground, and chunks of snow lodge on branches, on logs, and on fungus that grow on the tree trunks. The streams are winding ice-blue paths through cold winter silence.

On weekdays Howard was much too busy to cruise this piece of timber. But one Sunday before much snow had fallen, we drove as far as we could down an old logging road. From there we struck out into the timber and followed a trail that to me was hardly distinguishable, for underbrush had crept over it and the seedlings were growing fast. Howard hiked along as if he were leading me down Michigan Avenue. This trail led us past an old logging camp of John Schroeder's. The roofs had fallen in on two of the buildings, but the log walls stood as erect as if they had been built only months before.

Beyond the logged-over timber around the old camp site the tall trees closed in around us, and the trail became more difficult to follow. In one place Howard pointed out a wagon wheel protruding through the snow and surrounding a maple sapling. "This tree," Howard laid his steel rule against its trunk, "has been growing between the spokes of that wheel for a long time. See, it measures nearly five inches in diameter now. I first saw it on a trip this way with Dad. There's some discarded rigging, broken-down wagons and sleds, chains, and cables scattered around here, too."

"Who left this stuff here? Loggers?"

"No. This section's never been logged. Prospectors located here looking for gold. They found some, but not enough to get rich. There's more silver in these hills than gold, they say." The trail we followed curved around two tall pines and along a clear, sparkling creek.

"This is Gold Mine Creek, Julie. Over there stands a chute. That black hulk is a blacksmith's forge. There's an anvil, pans, and other mining gear there, too."

I looked in the direction Howard pointed toward. I saw, also, the familiar shape of a cookstove leaned against a tree. Its exposed parts were rusted and falling apart. Vines and weeds had grown up and surrounded it. Now it was partly covered with several inches of drifted snow.

"This lonesome place is known to loggers and deer hunters as 'Old Salt Pork Mine on Gold Mine Creek at the end of Gold Mine Trail.'"

"It's certainly a wild place." A porcupine scuttled up a tree, and the squirrels chattered their disapproval of our presence. Names occasionally mentioned and broken-down equipment were all that remained in the wilderness to tell the story of someone's wild dreams of sudden, easy riches. . . .

We left Gold Mine Trail and turned directly west into the sun. Howard found the markers he was looking for and began pacing rods and estimating the timber. I tramped behind him in silence while he counted trees and recorded the number and different kinds he found. When Howard was satisfied that he had checked enough of the area for his purposes, we left the virgin timber.

"Dad logged bird's-eye maple along this trail one winter. He worked for two months and found only enough bird's-eye to make three carloads. At a hundred and forty dollars a thousand, the logs he cut brought a good profit," Howard recalled a bit wistfully.

Because I had always wondered about bird's-eye maple, I asked, "How do you distinguish a bird's-eye maple tree from an ordinary maple, Howard?"

"Well, to men experienced in looking for bird's-eye trees," he explained, "they just *look* different, Julie. Their foliage is thicker, and their branches are very gnarled. When they are tapped with an ax eight or ten feet up on the trunks, they produce a hollow sound. Two men look for bird's-eye trees together. The lightweight man climbs up on the other's shoulders to sound the trees high on the trunk. Only expert, experienced cruisers can sight and sound a bird's-eye maple. Most loggers cut out a core of wood to be positive. Growing bird's-eye trees look much the same as ordinary maple; but felled, they show the 'eyes' that distinguish them as what cruisers call 'bastard trees' because their seeds are not fertile and do not reproduce. Just what causes this imperfection of eyes in maple timber is not known. Neither can anyone say why their seeds are sterile."

With this knowledge, I can no longer be hoodwinked by Steve or any other jack telling me about birds causing the eyes in bird's-eye maple trees.

That Sunday after I had followed Howard through rough wilderness country, straddling logs and struggling up steep gullies, I climbed into bed, aching in every muscle. Howard, too, showed that he was tired.

"That country is too rough to log I've decided, Julie. I've got a line on a piece of Keweenaw stuff north of Connorville. I think I'll write for an option on that so when we're through logging here I'll have some timber lined up."

Another Sunday some weeks later, Howard toted Bill Weston's supplies and cruising gear into the Bass Lake country with the tractor. Bill was a logger like Howard, but he spent much of his time cruising timber much as some men pursue hobbies related to their work. Bill spent many weeks camping, sleeping in a tent and tramping all day in all kinds of country to estimate the value of timber holdings. The hardships we endured living in a logging camp were nothing compared to the hardships suffered by all cruisers.

The only known winter campers in this region are the cruisers who find winter camping necessary because of the nature of their jobs. They solve their problems of choosing a camp site—of combatting winter elements, animals, and other hazards of the deep forest—by carrying with them warm bedrolls, extra warm and lightweight clothes, canvas windbreakers, concentrated foods, canned heat, a compass, and a gun. Without this equipment, they would not venture beyond the logging road least remote. These cruisers, lonely winter campers, are well versed in the lore of the woods, and they know how to meet nearly every emergency of winter camping—if not by experience, by tales told by other cruisers.

A week later Howard went back to get Bill.

"I was darn glad to hear your tractor, Howard. I was down to my last rations. Of course I knew you'd show up—afoot or by tractor. After eating out of cans, a good big hot meal will taste good again."

While we all sat around the camp table, Bill gorged himself with camp fare and told us interesting phases of marketing bird's-eye veneer which he had hoped to find on this trip.

Big money is paid for veneer sold for export, because veneer sent to foreign ports has to be expertly handled. It's peeled on the landing, and it has to be loaded without the use of tongs. The chains and skids used to roll the logs onto the cars have to be carefully wrapped with gunny sacks to prevent cuts and gouges from marring the surface of the valuable bird's-eye. The logs, cut no less than seven feet in length, are stamped with a letter and a number as they are loaded. For instance, a log stamped "A-1" would indicate that log as the first log loaded into car *A*. As the logs are shipped by railroad to New York, and then reloaded and shipped by water to be delivered to European ports, an accurate check can be kept of each log in the whole shipment.

"Bird's-eye maple, now a scarce and valuable timber product, was sorted out of hardwood flooring logs and culled years ago. Today," Bill told us, "through a French syndicate with representatives working in the United States, I shipped carloads of bird's-eye to Havre, France, and Genoa, Italy. Once I went with a shipment to Havre. The veneer was sold in small chunks, three to five feet in length, to cabinetmakers. When the ship docked, this blond wood, which is so scarce in France and Italy and prized so much for making fine hand-carved and inlaid furniture, was unloaded and sold right there on the wharf. Cabinetmakers came with their donkey-drawn carts to bargain shrewdly with the agent for the piece of bird's-eye they wanted. When their buying was finally completed, they loaded their precious wood carefully into their two-wheeled carts and went away, to return, I was told, the next time a ship bringing choice American veneer docked.

"I don't know what the cabinetmakers pay for their small pieces, but I've gotten as high as three hundred dollars for a thousand feet of choice bird's-eye. That sounds like a fabulous price, but the reason for that is because there is so little bird's-eye timber left. Most of what is left can be found in northern Michigan and in some parts of Pennsylvania."

Howard and I visited Bill's logging camp one Sunday. Bill had some estimates on the timber north of Wakefield that Howard was interested in seeing. Bill's camp buildings stood at the edge of a clearing near Ethelwood Landing where Howard's dad had last logged. In a wide clearing, crowding his camps, stood impressive decks of logs covered with snow.

We walked along a wide, elevated boardwalk that connected the bunkhouse and the cookshanty with the office. I was curious about the reason for the walk, and asked, "What's the reason for this elevated walk, Bill?"

French Cabinet Makers Came with Carts

"Well, it saves shoveling snow all winter for one thing, and in spring the jacks can't drag much mud in the buildings. It falls through the wide cracks."

Bill's office was equipped with the usual counter, desk, high stool, barrel stove, several small-paned windows, and a bench or two. His sleeping quarters just off the office were well furnished, I noticed. A long shelf of books, an easy chair, a large table, and a big, comfortable-looking bed with a thick Hudson Bay blanket rolled at the foot indicated that, once here, Bill stayed for long periods of time. Maps of most of Michigan, Wisconsin, Minnesota, and much of Canada adorned the walls.

While I waited for Bill and Howard to finish their business, a lumberjack suddenly pushed open the office door and lunged into the office. Obviously, he had just returned from town. He walked unsteadily and talked loudly, using language that lumberjacks know and use continually among themselves, I am told, but refrain from using in any woman's presence. He began an argument with Bill about why he couldn't get something to eat. Exactly what he wanted was, "a dang good meal, vit spuds. Hot, too. None a dis here cold stuff, see."

Bill explained, patiently at first, "It's just not logging-camp policy to expect the cook to have *hot* meals ready to serve at all hours, especially on Sundays. And you know that as well as I do, Jack."

The jack began to use more curses and obscene language to argue his point. I saw Bill's face flush and his fists clinch. Bill was a bull of a man, a Paul Bunyan of a logger—well built, muscular, and capable of strenuous work and arduous cruising. His black eyes flashed, and he ran his big hands through his graying dark hair. "Better get out, Jack." Bill's voice was thick with anger and impatience.

Defiantly, the jack refused to go.

Bill, red-faced and trembling from anger, put his big hands on the counter, and, with a mighty leap, he vaulted directly over the high counter. He landed behind the surprised jack with a heavy thud. Bill kicked the door open with determination; and before it had swung back, he lifted the obstreperous, intoxicated jack by the seat of his pants and the collar of his shirt and heaved him outside. The jack landed sprawled on the boardwalk.

Bill threw the jack out so quickly that neither Howard nor I had time to do more than change our expressions from wonder to amazement at Bill's strength and adequate method of shutting up this loud-mouthed lumberjack.

As Bill hooked the door, he said, "He should know better than to talk that way in here." This was offered in the nature of an apology for having thrown the jack out, but Bill's actions were to us understandable. I am sure anyone would be irritated to be forced to submit calmly to being called all manner of names by a drunken jack. However, in his sober moments, I doubt if this jack would have acted as he did.

The sudden shock of being thrown out onto the walk jarred the jack into sensibility. I watched him get up, shake his head, and, muttering to himself, go off into the bunkhouse. I suppose he slept all the rest of the afternoon, evening, and through the night, not remembering in the morning anything about the whole affair.

Howard and I returned to Tula Camp late that afternoon to find that our impregnable bunkhouse had been suddenly invaded by a *girl!* No mousecapades in a girls' dormitory could compare with this episode.

Ed had perched himself where he could see clearly the bunkhouse door. With a twinkle in his eye he told us, "A gal just came in sellin' magazine subscriptions. She worked on me a while. Then I told her ta see the guys in the bunkhouse. When I said there was fifty lonesome guys down there that just loved ta read, she high-tailed it down there in a jiffy. I'm a waitin' ta see how many subscriptions she sells them dopes."

I heard the door slam shut behind the girl who had discovered a way to contact many potential subscribers with little effort.

"This is gonna be good," chuckled Ed. Howard and I crowded behind him to watch, too. This brave girl was invading a place where no one, except the boss, dared intrude. Females were strictly taboo.

The lumberjacks responded to this invasion of their bunkhouse like sailors puttin' ashore off a slow boat from Russia. They piled out of the front door in swarms. Some made a beeline to the cookshanty for, I suppose, coffee and doughnuts. The teamsters headed for the barn, perhaps remembering that the horses, already fed, could stand a few more oats. Charlie made a dash for the tractor shed to be sure the tractor was greased and to clean out more twigs and branches from its tracks.

Embarrassed jacks poured out into the sunshine. Some went out the back door down the private path. When the traffic became too heavy in that direction, the others were forced to take to the timber. A few came up to the office for protection. A lazy Sunday afternoon for the jacks had turned into a busy one, in no time at all. The camp was alive with bewildered lumberjacks; some appeared in their stocking-feet, others came out without shirts, some bolted out the doorway, hastily fastening their suspenders to their pants. All were desperately trying to escape the "mouse." Several very stubborn or brave jacks stayed in the bunkhouse and became subscribers to different magazines. We enjoyed one for five years, because, in the spring, the jack who bravely signed the subscription blank left no forwarding address and did not return to our camp again. Thus, this winter Sunday became not only eventful, but profitable as well.

The girl left Tula Camp, never knowing what a social blunder, according to logging-camp customs, she had committed. Neither did she know what a turmoil she had caused. I wondered at her supreme audacity.

During that evening in the glow of soft lamplight we

chuckled over the excitement, and Howard recalled a visit from a peddler and watch-repair man.

"A couple of years ago, Julie, a fellow made the rounds of all the logging camps. He sold, as he said, solid gold watches. This peddler wasn't an entertainer, but he was a darn good auctioneer. He walked into camp just at suppertime. After he'd eaten he began talking the jacks into trading their old watches for his with their gold cases. Some gullible jacks did select new watches and traded in their old ones. Claiming that he had no use for their old watches, the peddler auctioned them back to the jacks. The auction jabber was the most amusing part of the transactions. Some jacks bought back their watches for only a few dollars.

"As I remember, this peddler also collected watches to repair and clean. He promised to return soon with them all in good condition. Because he gave carefully written receipts for all the watches, the jacks thought he was honest.

"He stayed two nights to transact all his business. The next night he sat in on the jacks' friendly poker game."

Ed interrupted Howard's story. "Yeh, the old skinflint said he didn't know a darn thing about poker and guessed the jacks could teach him a few pointers. B'fore they got through that peddler taught the jacks a few things about not trustin' every gol-dang feller that comes along. Skun 'em, that's what he did. The jacks played all night tryin' ta win their dough back, but b'fore they knew it, it was daylight and time ta turn out ta the timber. When the jacks come in at noon for chuck, the peddler was gone—watches, poker money, and all. Was them jacks mad! They cussed and threatened to get even with him. The peddler never did come back with their watches he promised ta repair, and the ones in them solid gold cases quit a-keepin' time in a few weeks. That made the jacks madder yet. Some jacks was lucky and bought their old watches back from the fellows who'd got 'em in the auction.

"All winter them jacks fermented up a dang good hate for that gyp peddler. Jacks don't mind givin' their money away, but they sure hate ta be gypped by a skinflint. I heard that them jacks saw that there peddler in spring breakup some place in Hurley. With assault and battery in their eyes, them jacks searched the town from one end ta the other. That gyp peddler knew what was good for him and vamosed. If the jacks had got him b'fore they got drunk, there'd a been as bloody a fight as you'd a ever seen."

"There were many kinds of camp visitors during the years when jacks stayed all winter at one camp," Howard said. "Dad used to have preachers, trappers, and salesmen of all kinds make regular trips year after year to his camps. The jacks were always glad to have them come, for their visits broke the monotony of the long winter months. That girl who came today sure broke the monotony for our jacks. They'll never stop talking about this visit!"

Then Howard recalled times he had spent with his father whom I knew he missed very much. "Many winter weekends I'd spend in camp with Dad. Mom would button me into my warmest clothes, and Dad would swing me into the cutter and tuck the fur robe tightly around us. Sometimes my cousin Apie would come along. We'd wave good-bye to Mom and be off. The horses' hoofs threw back clumps of snow, and the harnesses jingled a merry tune."

"Was it far to your father's camps?"

"The ride was long enough to give us cold hands and feet; and if we weren't hungry at home, out at camp we ate like we'd been starved. To tell the truth, we were starved for the food served in loggin' camps. I can still hear Dad sayin', 'Boys, you eat so gol-darn much I'd a whole lot rather pay yer board than feed ya.'"

"What could boys do after supper in a logging camp? I suppose you got sent to your bunks early."

"No, Dad was a good sport and let us stay up later than Mom ever did. After we'd dried the knives, forks, and spoons for the cookee by swooshin' 'em around in flour sacks, makin' a big clatter, we visited the jacks in the bunkhouse to listen to a wild tale or two. We were glad when we heard a few swear words we could repeat to shock our mothers' friends when the Missionary Society met. Or we'd go the rounds with the barn boss, fightin' about who'd carry the lantern until he was forced to get two lanterns. In their pale, insufficient light we helped the barn boss fill the oats boxes, make the halters secure, and shovel out the manure. When we returned to the office our clothes smelled like the barn."

"Didn't you ever get tired of camp life? Of course you didn't or you wouldn't be logging. But wasn't there some kind of amusement besides reading and talking for the jacks?"

"Not much. I remember one time a traveling preacher came to camp and gave a loud and long sermon. Then there was the peddler of knives, watches, and trinket jewelry who hit the bunkhouse one night when I was there. He sold trinkets and watches but primarily he peddled entertainment for the jacks. He came back year after year, Dad said, for he evidently enjoyed his unusual occupation. He arrived late in the afternoon, laden with his wares, and his squeeze-box strapped across his back. His bright red tassel cap bobbin' up the trail announced a rough time in the bunkhouse that night.

"As soon as the jacks came back from the cookshanty after supper, they settled themselves around the peddler in the yellow glow of the hanging kerosene lamp to listen to his stories, his catchy accordion music, and his lusty songs sung to his own accompaniment and the throaty hummin' of the lonesome jacks. He gave a one-man entertainment which, judged by lumberjacks' standards, couldn't be beat. I didn't care for the songs. Dad liked the accordion music best. We both were dumfounded by the peddler's sleight-of-hand tricks by which he made the handkerchiefs and coins of accommodating jacks disappear. When one jack grumbled too much about losin' a two-bit piece, the peddler found it in the jack's own trouser cuff. That sure got a big laugh out of the jacks. They slapped their knees, guffawed, and howled the rest of the evening at even the stalest of the peddler's jokes.

"In this atmosphere of merrymakin', the peddler opened his cases of fancy knives and jewelry and hand-engraved silver watches. With glib words and persuasive salesmanship, he told of their excellent qualities. He compared his engraved watches with their plain ones and proved how sharp his knives were by shaving a thick sliver from the stove poker. The jacks bought his wares just to repay the peddler for an evening of fun. Then, one by one, the jacks, unwilling to leave the congenial group but unable to stay awake, hit the hay. One boot hit the floor and then another, until the peddler, too, put away his squeeze-box and climbed into an empty bunk at the end of the bunkhouse to snore away what remained of the cold winter night. In the darkness of early mornin', after eatin' a hearty breakfast, the peddler hit the trail to another camp, singin' with gusto. He departed with the good will of all the jacks who guffawed over his tales, sang his songs of many verses, and tried for days to figure out his slick handlin' of their coins and handkerchiefs.

"Back in the office my cousin Apie and I climbed up into the upper bunk after peelin' off the top layers of our clothes down to our itchy underwear. Above the top bunk the heat hung in smotherin' layers while the barrel stove was red-hot; but before mornin', with the fire out and the stove black and cold, we'd curl close together and cover our noses to keep them from freezin'.

"In the mornin' one or both of us would come sailin' down, clutchin' a blanket, strugglin' to land where it would hurt least, to find it was no use. We'd hit the hard, bare floor and it always hurt. That was part of the fun of comin' to camp—those fights to see who was the better man. Dad let us scrap plenty. The tussles in the upper bunk were as endless as our trips to camp."

I could have listened to Howard's storytelling far into the night, but Ed would remind us that it was time to hit the hay by winding his watch and turning down his blankets. As a final reminder of the late hour he would noisily fill the stove and turn the damper.

Weekdays passed rapidly, and quiet Sundays followed. One memorable Sunday, Howard and a good friend of ours, who is an expert electrician, drove to Silver City where they appraised and bought a light plant. The next night they installed it and wired the camp buildings for electric lights. Until then, we had used kerosene and gas lamps during the long winter evenings. We dressed by lamplight, ate breakfast and supper by lamplight. We sat around talking or playing cards by lamplight; we undressed and went to bed by lamplight. No one strongly objected, (unless the bull cook did when we couldn't hear him—it was up to him to keep the office lamps filled and cleaned). What reading and sketching I did, I had preferred to do during the day. I had found it inconvenient to do the bookkeeping at night, because for me to do figuring I have to have quiet and freedom to exclaim when and as I feel moved. During the day, alone in the office, my exclamations went unobserved.

The real value of electric lights for me was the ease with which they were lighted. Howard arranged the wiring so that all we needed to do to start the light-plant motor was press the button on a switch in the office, and the motor started—usually.

I could light a kerosene lamp easily. But lighting a Coleman lamp was quite a different matter; and rather than fool around with them, I had always used a kerosene lamp. The directions for lighting a gas lamp must be learned by heart and followed to the letter. The fragility of the mantle, a gauze-like, conical-shaped thing, hung on a frame inside the tall slender chimney, constitutes a mental hazard for me. I am sure I ruined more of these mantles by just thinking of their delicate nature than by actually touching them.

The length of time it took for our gaslights to become as bright as possible constituted about a quarter of the whole evening. If, by accident, the lamp got turned too high, the whole mantle burned black; and, temporarily, while the accumulated black burned off over a low light, we sat grumbling in light inadequate for distinguishing even familiar faces. Between waiting for the gaslight to become bright enough or the mantle to become free of carbon, most of our evenings were spent in semidarkness—comfortable light, but not too satisfactory.

When I had wanted lots of light, I lighted all the kerosene lamps I could find in the office and lined them along the counter. Of course, I prefer electric lights to every other kind except candlelight and the glowing light from our own cheery fireside; but for sentimental reasons I have kept a kerosene lamp, originally my grandmother's, which cast its gentle light on my mother's and father's supper table when they were first married and lived on a Kincaid claim in Nebraska. My excuse to my friends for having such an antiquated thing around has been, "Oh, it's handy when the electricity is off." It surely is.

At Tula Camp the electric lights went off frequently, and Howard had to tune up the motor before we could put away the kerosene lamps again.

The electric lights meant most to the bull cook, cookee, and the barn boss. The cookee cleaned all the lamps used in the cookshanty, the bull cook cleaned those used in the bunkhouse. For them, washing and polishing lamp chimneys was conducive to the cultivation of a long list of profane words. While Squint scrubbed the floor and the odor of soapsuds and boiling water merged with kerosene smell, I ate my breakfast and listened to the cookee grumble and mutter every Saturday morning over a row of blackened chimneys and empty lamps. Our electric lights at camp eliminated this disagreeable task from the many duties of our bull cooks and camp cookees. The barn boss now had light enough on dark winter mornings and evenings to make it easy for him to keep track of all the horses' heels at once.

The jacks enjoyed the added convenience of electric lights in the bunkhouse and the cookshanty. They found mealtime more pleasant and discovered that it was more difficult to slip a pair of aces out of one's sleeve in the glare of the light bulb hanging over the poker table. Steve said, "Since we got electricity, my poker luck's sure changed some." For better or worse, he never said.

Howard, the last one of us to bed, would give us all a warning signal by flashing the lights off and on again. After a short time he would flick the switch off, and starlit darkness of the cold winter nights would envelop Tula Camp.

One such night, long after the lights had been flicked out, I sat up in bed as suddenly as I had gone to sleep. What woke me I did not know, but I was shaking with fright. Everything was quiet except Tex under our bed, whining a little in his sleep. There were times when I wished he were not there, because our bed dipped low and near to the floor. When he changed positions, he bumped the mattress and many times woke me from a sound sleep. This night I was glad to have Tex under my bed, because I felt he had not awakened me and, if need be, he would offer protection.

Strange complete silence challenged my instinct. Not even the usual sounds of horses stamping in the barn came to me as I listened for an indication of what had roused me. A burned-out log that collapsed in the stove startled me.

Then, from the timber came a weird, crazy scream! It was a scream that sent chills down my spine and ruffled the hair on my arms. I shuddered in horror under the warm covers. A black shadow flicked across the patch of moonlight on the floor. Only the wind blowing the branches of the saplings outside, I hoped. The next second seemed like a century.

The wailing scream sounded again! I sat up in bed. Whatever made that unearthly noise was closer now, it seemed.

Tex had heard the scream, too. He sprawled flat to make his way from under our bed. He woke Howard in passing under him. Again the scream sounded through the winter stillness. The cry was a most horrible wail—the scream, of a crazy woman right under our window. I tried to speak and couldn't.

Tex yipped, pushed his nose in the crack of the door, opened it, and squeezed out the doorway. The door slammed shut. At last I could whisper. "What was that unearthly sound, Howard" My voice quavered and revealed my nervousness.

"Wildcat. Haven't heard one for years. Sure is a terrifying sound. Probably smells the horses. He won't stay around here." In a moment Howard was snoring again.

Twice more I heard the screaming wildcat, but after I knew what the bloodcurdling noise was, it no longer had the same terrifying effect. Still it was not the sweetest of sounds to me. That haunting wildcat's scream remains the most frightening noise I have ever heard. Hours passed before I went back to sleep. My thoughts were of Arvid, the crippled shacker at the far side of our timber, alone by choice, with not even a dog to answer that eerie cry. After the barn boss, on his way to feed the horses, slammed the bunkhouse door and a pink tinge of morning showed through our window, things seemed normal again. I relaxed, to be engulfed by sleepfulness, a deep sense of security, and the warmth of the blankets.

Fellers of Our Timber

Forest trails and logging roads are the special thoroughfare of lumberjacks—the fellers of timber whom Howard has known all his life. Season after season I watched the jacks as they walked alone or by twos in the sun and snow, rain and mud into Tula Camp and along the highways. Strapped to their backs they carried their packsacks, filled with all they possessed. Some were pathetically empty, Howard told me.

"They tote around unwashed underwear, patched and thin; a faded chambray shirt or a threadbare woolen one; felt shoepacks and woolen socks, smelly and stiff with sweat; pamphlets advertising luck charms and sex literature; a comb; and perhaps a razor. Sometimes they carry a torn and faded picture of an old-country girl or a picture of a family faintly resembling the jack in whose packsack it is found."

Lumberjacks climbed the hills of our logging roads, rested on the stumps in the sun, silently came into camp and talked to Howard the boss, or Ed the foreman. They worked for a week—or a month—or stayed until breakup—and silently left. They took with them varying amounts of money—money earned in one of the most hazardous of occupations. These men "pulled the pin" and hiked down the road without friends or relatives waiting for them. They searched out what companionship they could find. Homeless and weary, they entered where shelter and warmth—physical and emotional—could be had in exchange for their hard-earned dollars.

Jacks Taking "Fire Time"

There is an end to the winding forest trail trod by the lumberjacks who spend their days in endless toiling and their freedom ingloriously supporting tavern keepers, hock-shops, women of the street, and their own companions of the logging camps. Jacks do not, of course, live forever. Some die natural deaths. Some are killed in woods accidents, and others take their own lives. Inexorable death finds lumberjacks easy victims. Physically spent from hard work, sick from drinking cheap whisky, and thinly clad in patched clothes, a lumberjack drops in a snowbank or sleeps under a tree during a fall of chilling rain. He fails to make camp. Only the tall trees witness his death.

Some jacks die in the white snow outside their shacks, overcome by age and work, to be found frigid with cold and death by hunters or loggers. A porcupine may have gnawed off part of the body before it is discovered. Others die in their shacks and are found in bed, death having come after a hard day's work. Howard brought one jack out of the woods on a sleigh in the middle of the winter—a lumberjack who had died in his bunk at Louie Anderson's camp. Another jack died of severe burns that consumed his shack. Other jacks die violent deaths of their own premeditation. Having no families, they live lonely lives and become morose and churlish. Some become so morbid and dwell upon their misfortunes at such great length that plans for their own destruction formulate. When jacks commit suicide, are accidentally killed, or die of natural causes, the newspapers print their death notices, and the circumstances of their deaths. Those who have no families or known relatives in this country are buried after services in a funeral home. They are the quickly forgotten.

—One evening Howard could not find the stout starting rope for the light plant. When it could not be found any place, Howard cut a piece from my clothesline and used that instead. The disappearance of the starting rope was not thought of until a week or so later.

The men filed into the cookshanty as usual for breakfast one morning. They took their accustomed places. One place, that of the teamster Pete, was vacant. For Pete to miss breakfast was unusual; the teamsters got up earlier than the other jacks to feed the horses. Such chores gave them several hours to work up a good appetite.

Howard finished his breakfast first and came back to the office. "Pete didn't come in to eat this morning. I sent the other teamster to look for him. He's probably sound asleep in his bunk."

I finished dressing. The office door swung open, and Joe, pale and shaking, burst through the doorway. "He's done it, he's done it!" Joe blurted out hysterically. "He said he would, now he's done it."

"Done what?"

"Hung his self—in the barn. He wasn't in his bunk—stuff all there just like always. Packsack under his bunk, and his hat still a-hangin' there. I thought maybe, just maybe, he'd got kicked in the head after I left him. He wasn't through with his jobs when I was, so I cum back ta the bunkhouse a'fore him. When I got ta the barn jist now, the doors was hooked. 'That's funny,' sez I to me. I spied a shovel by the manure pile and went ta prying the door ta the hay barn open. 'Twas wired shut, it was. When I gits it open, there—a-hangin' and a-swingin'—is Pete. I tell ya he's hung his self."

Howard pushed his way through the men who had gathered in the office. I threw a jacket around my shoulders and followed. We ran to the barn, where bales of hay and sacks of oats were piled. There, from a beam of the low barn, hung the body of the teamster, ghastly in the early morning light, pathetic, limp, and inert. His head drooped, and his eyes bulged. The doorway filled with the forms of men gazing upon the incredible sight of Pete's body turning slightly in the noose so recently fashioned and secured to the beam just above and beyond the pile of baled hay. He had evidently planned his death well, and, not being disturbed, had consummated his own destruction by taking one big step from the hay pile into space. Now it remained for someone to cut the rope.

"He's dead. We'd best let the coroner handle this." Howard's face was chalk white. His words were barely audible.

Howard closed the barn doors. We all went back to the office. Howard found it useless to send the crew out as usual. The jacks lingered in the office reconstructing the events preceding this morning. Howard drove to Wakefield and returned with the coroner, the State Police, and an ambulance. After the coroner inspected the body, a trooper cut the rope and lowered the body to the waiting stretcher. From my window I watched the men cover Pete's limp form with a gray blanket and carry it up the road. They slid the stretcher into the ambulance and closed the doors. Axel ran out of the bunkhouse with Pete's packsack.

The bull cook came in for clean blankets to put on Pete's vacant bunk. For some queer reason, he left the bunkhouse door propped wide open all day.

With great reluctance, Joe and the other teamsters went to the barn to harness their teams. Late in the morning the jacks finally left for the woods to try to do their respective jobs. They found it no ordinary event to discover that a fellow-jack had hanged himself in the early morning while they ate. All the men seemed greatly shocked. Some even refused to go anywhere near the barn. Many weeks passed before the men and Howard and I stopped talking about how Pete had brooded and mumbled to himself about how useless he thought it was for him to go on working when he felt sick. He had continually brooded about how old he was and that he had no one to care for him.

"He told me a good many times, 'I'd be better dead.'" declared Ed.

Then the rope from the light plant had disappeared. That short length of rope had broken Pete's neck and ended, horribly and quickly, his sick thinking. A collection taken up among the lumberjacks, added to the wages he had coming, paid his funeral expenses. He was quickly forgotten by all, evidently, but not by Howard and me.

Jacks may not have what you or I consider the best of morals. Their language is rough. Most of them are not clean. They are generally spendthrifts, using their money to buy whisky and women. But they work hard. They are not hypocrites, pretending to be what they are not—which, in my humble opinion, is a commendable virtue. They appear to be tough characters but are really harmless and goodhearted. Among all those who came and went from our camps, none ever treated me with disrespect. A most abject-looking jack was Emil. He did not know that I was in the office and came in from the barn, roundly cussing Kaiser for giving him a smart kick on the knee. Among themselves they may have used vile language and expressed base comparisons, but I heard only cursing. Out of fairness to the jacks, I let them know I was around by singing or whistling well in advance of my appearance, to save them embarrassment.

I knew no jacks who could claim to be meticulously clean. Almost all were inclined to abandon shaving, bathing, and washing their clothes until they made a stake. Then they bought new clothing, got a shave and a haircut, and had a drink or two. Some called on their lady friends then.

A few filthy jacks found their way to Tula Camp. Howard fired two the day after they had been hired. Generally, our crew took care of other undesirable jacks in their own ways. The men crowded the offenders at the table, caused their wet socks to fall off the drying line, and made things so uncomfortable for them in many other ways that the unclean jacks moved on of their own accord.

"We sure deviled that guy with 'lodgers' plenty," Steve told me, as we watched a jack with lice hit the trail. "Took the hint, too. Didn't think he was a goin' to till we broke that raw egg in his boot."

When jacks have spent all their earnings, they sometimes

borrow from their friends, from bartenders, or from storekeepers. I have been told they pay their debts promptly when they hit town again.

"In hard times the jacks became professional panhandlers," Howard told me. "But today there are only a few who ask strangers for loans."

Occasionally, jacks who have money coming from Howard ask me for a dollar or two for drinks; but I learned to refuse, because when I did give them money they continued to come back for more, each time a little more talkative, shaky, and demanding.

About their pasts jacks are silent and uncomplaining—unless they have been drinking. They resent attempts from outsiders to pry into their personal affairs, even if the prying is done in a sympathetic manner for their own good.

One day, among a number of lumberjacks going to Hurley who managed the entrance to the bus with uncertain steps and sank into the nearest seats, I saw one who stayed apart from the others. He found an empty seat ahead of me and huddled close to the window. He gazed out with unconcern for all around him. Some of the jacks loudly made known their affairs to strangers in monologues, complete with gestures. For this particular jack the interval of the bus ride to Hurley was managed in utter silence and borne stoically.

While the bus filled with people, a woman found the seat beside this jack ahead of me the only vacant place. She settled down comfortably and began a conversation which consisted only of questions. At first I thought the lumberjack did not hear, as he seemed to be paying no attention.

The lady asked, "Are you married? Do you save any of your money? Where are you going? Do you ever go to the movies?" Finally, without getting an answer to any of these questions, she asked, "Where is your home?"

If these questions were just casual conversation to her, I think the jack considered them malicious prying and none of her affair.

She persisted in questioning him. At last the jack became exasperated to the point where I could tell that he was "damn mad." He sat upright with a start, grabbed the brim of his battered hat in both hands. He jerked it down on his head with significant force.

"Lady," he growled from under his hat, "when me hat's on me head, me family's under shelter."

With that outburst to quiet her, the lady indignantly flounced to the front of the bus and stood until she reached her street.

A number of young men working for Howard, the loading crews and the tractor and truck drivers, were not typical homeless lumberjacks. These fellows lived in neighboring towns. A few of the sawyers who worked at Tula Camp were men who farmed during the summer in Wisconsin and Michigan. These farmers came in the fall as soon as their chores for the season were finished. Aside from going home for Christmas, they stayed in camp until spring. Then, of course, the plowing and planting of their farmlands needed to be done, and they felt compelled to go. Their winter sawing for Howard was a thrifty way for them to make two seasonal occupations provide an ample year-round income. These men were dependable, excellent woodsmen who were always on the job early Monday morning when the other jacks were recovering from trips to Hurley.

The majority of Howard's jacks were, however, men without wives or families. The lack of responsibility to anyone but themselves made it easy for them to pick up their packsacks and leave camp at will. When they moved from one camp to another too frequently, they became known as "camp inspectors." Camp inspectors moved for no apparent reason other than their desire to try the chuck, look the timber, and make numerous comparisons in the bunkhouses of all the surrounding camps. Some camp inspectors stayed at Tula Camp just overnight; others lingered a few days, according to whether they liked the chuck or whether our men were friendly toward them. One morning after breakfast they would leave without notice. Howard considered himself lucky if the camp inspectors cut enough logs to pay their board.

Howard tried to make every other Saturday payday. However, the men could draw money against their accounts any time they wanted to. Every two weeks the saw gangs' strips were counted and their accounts figured. The time put in by the men who worked by the hour was figured the same week.

But some jacks considered any day they wanted to go to town a payday. When one sawyer of a saw gang quit, his partner was forced to leave, too, to find another jack with whom to work. If sawyers quit and came in for their checks, their jointly cut strip was counted as soon as Ed found the time. Then Howard would pay them off. The jacks' thin excuses for wanting to quit or go to town for the weekend were never questioned, and they avoided any discussion of the real reason. Silently they would fold their checks, put them in their inside shirt pockets for safekeeping, and leave with their eyes downcast. If they intended to return on Monday morning, they left their packsacks in the bunkhouse; otherwise, they hoisted their packsacks to their shoulders, walked down the road, and seldom came back. Sometimes a satisfied sawyer returned with a new partner.

When jacks quit, I thought they were unfair to Howard. But he had a tolerant explanation. "Timber is not all good cutting chance, Julie. Some strips are exceptionally thickly timbered, with many logs in one tree; on other strips the trees are farther apart and small or unsound. Occasionally a logger finds that it's impossible to lay out strips of timber with equal amounts of good and poor cutting chances. If sawyers come to a strip that does not look good to them, they usually leave; and their reason is obvious. Most good sawyers know that, if the cutting prices are right, they can make up on a good strip for hard work they've experienced on another strip. When it is certain that the cutting chance is too poor to allow a reasonable profit for a lumberjack, I raise the cutting price, or else leave the strip till it will pay out by the work of other sawyers.

"Some jacks prefer to go where the cutting is always best, but many are able to make good wages cutting all kinds of timber. These, of course, are the energetic jacks."

At Tula Camp the timber grew tall, straight, and large. The jacks who quit were never missed. New gangs of sawyers came up the hill to the office weekly to fill the vacant bunks of the jacks who left.

Most Saturdays, before their checks were all written, the lumberjacks would begin gathering outside the office and around the stove inside. I would listen to them make arrangements for rides to town with the truck drivers. Several would state their preference for walking, because they hoped to get a ride all the way to Hurley once they reached the highway. The atmosphere was very different from that when men actually quit. These jacks were just taking a spell of lay-off and going to town, well heeled and ready for a spree minus the bother of taking packsacks.

They kidded each other about what each thought the other meant to do when they got Hurley. They mentioned redheads and blondes—some by name—indicating that these girls probably excelled in their profession. From this talk, I gathered that the destination of the jacks was not the department stores to buy shirts or pants—the excuses they often gave me when they thought they ought to explain why they needed money on a certain Saturday morning. All would go their separate ways, or two jacks would stay together for company and potential protection from oversolicitous girls and saloon keepers; but they would all see one another at some time on Silver Street.

Some who walked found it convenient to stop at Bingo's Tavern, where I occasionally went for a case of pop. There they cashed their checks and bought a round of drinks for all in the place. Warmed by their choice of amber liquids, the jacks started walking down the highway again, plodding slowly to their destinations in Hurley or Wakefield.

By late Sunday afternoon most of the jacks had spent their money. Those jacks who were not staying on to bum further drinks began their devious travels back. Sometimes it would be the camps they had come from. Sometimes it would be some camp about which they had heard rumors of better chuck, bigger pay, or taller timber.

Loggers who found lumberjacks hard to hire have been known to employ "man-catchers"—usually someone connected with the police force in Hurley—to highjack drunken jacks, take them in their partly unconscious conditions to their camps. The price per head ranged from five to fifteen dollars, according to the distance the mancatcher had to haul jacks. When the jacks sobered up, the loggers persuaded them to stick it out until the next payday. Penniless and hungry, the jacks usually decided to stay, which was exactly what the mancatcher and loggers expected—once the jacks were nabbed and hauled to camp.

Pete, Emil, Ole, Steve, Laggin', and most of our jacks did not allow themselves to be lured away from their jobs by mancatchers. They came back to Tula camp, at times bedraggled, with scratched faces, torn clothes, and black eyes from brawls they had been in. (The police as a rule in those days looked the other way when fights and arguments took place on Silver Street. Only when the situations became complicated and bloody did they interfere.)

On Monday mornings Howard's jacks were wobbly on their feet. They had little or no appetites but were quite willing to go out sawing and skidding to work off the jag.

Liquor Bar and Gambling Table

We, too, spent any number of Saturday or Sunday nights in Hurley. It was hardly in the same way, however, that Laggin', Steve, Eino, Ole, Tauno, and the other jacks did.

During the first year Howard and I were married, I went with him to Hurley for the first time. As we came out of the theater in Wakefield, dazed by the beautiful girls, elaborate interiors, and enchanting music of the show, Howard suggested, "Let's come back to earth and go some place to eat. I have to hire some sawyers. Do you want to drive to Hurley? We can have a steak or Chinese food at Doc's Place in the Burton House. It's not a classy restaurant, but I like the food Doc serves. Or how about going to the Romance Lounge in Ironwood?"

I could tell by the scorn in his voice when he mentioned the Romance Lounge that if I chose its chrome-and-leather interior, illuminated by ghastly fluorescent lighting, this might be the last time he would take me out to have a steak, or anything else, for that matter. Because Howard was my main interest now, I made my choice without hesitating a second.

"Burton House for me, thank you. Whatever Doc's Place turns out to be, I am satisfied just to be with you, Howard."

We left Wakefield and approached Silver Street by driving first through Ironwood, on down a hill, around a curve, and over a bridge spanning the Montreal River, the boundary line of Michigan and Wisconsin. From the bridge along Silver

Street in Hurley to the Burton House, a distance of six blocks, I counted seventy-eight taverns and liquor stores. It was their brilliant neon signs that created the crimson glow we had seen in the sky for miles before we entered the town.

Howard, who has picked up lumberjacks in every saloon in Hurley at one time or another, told me, "Some places have been renamed and redecorated many times, while others are much as they were when Hurley began to mushroom into a hell-raising, riproaring mining and logging town. Some (defying all efforts of the police to abolish gambling) still retain their gambling tables and mahogany back-bars." But they all sported huge and fancy neon signs, I noticed.

Howard parked on Silver Street. While I waited in the car for him to hire lumberjacks, I watched the jacks come and go aimlessly along the street, staggering into particular taverns, then emerging only to go into the nearest ones—next door or across the street. The names were colorful: Round Up Bar, The Red Feather, Hi Ho, Dell's Place, Tony's Pastime, Club Fiesta, and The Down Beat. (Recently, I might say, there have been added Pearl's Harbor, Victory Bar and 4-ever Amber.)

In and out of these places the jacks made their way, slowly at times and uncertainly. Then a burst of energy or a sudden idea would take them in a reasonably straight path to another plate-glass front, painted black on the inside to a line above eye-level. I was amused when I saw one lumberjack reach an entrance just as the door swung out, opened by a departing jack. Two not-so-steady lumberjacks tangled for a moment, one trying to go in, one trying to come out. The struggle ended with both jacks staggering inside, arm in arm, one to buy the other a drink, no doubt, and thereby untangle themselves.

I have learned that most old-time jacks had in common a ruthless disregard for money; what they worked so hard to earn meant nothing to them. They spent it all before they returned to camp; and if it was not all gone, they could not be persuaded to leave Hurley for the highest prices ever paid for cutting timber. A place to sleep, food, a few warm clothes, and liquor satisfied their material wants. Some wanted women; others scorned them. Accumulation of any possessions other than necessities impeded them in their treks from one camp to another and to Hurley. Their desire for amusement and entertainment was easily fulfilled. Silver Street, lined with saloons, gambling houses, and an occasional brothel, was their mecca, a reflection in general of what seemed to interest them most. I thought that night while I watched them (as I have thought many times since) that these men, who were exposed every day they worked to the majestic beauty of virgin timber, responded to that environment in a manner that was entirely without logical explanation. The clean snows of winter, the warming suns of summer, the delicate laciness of leafing trees in spring, and the brightness of cobalt skies and vivid foliage of autumn seemed to leave no mark. They remained uncouth.

Howard returned to the car and suggested that if I wanted to see the inside of a Hurley saloon I could safely go with him to have a glass of wine. Who would not be curious about the interior of what then constituted a brothel? All that appeared from the street was the painted plate-glass front. With some reluctance on my part, we went in, first through the swinging doors to the barroom. Light from the chrome and fluorescent fixtures on the ceiling gave Howard's face (mine, too, I suppose) a bilious look. On one side of the room straggled nondescript tables and chairs, and on the other there reared an ultramodern bar at which three customers stood in various abject positions. Cigarette butts, papers, and used matches littered the floor. On the sooty walls hung calendars displaying legs, torsos, breasts, arms, and beautiful faces of exquisite girls, which made the place a gallery of dimpled peach-blossom flesh.

An oil burner produced a smelly heat; blue smoke hung in thick clouds over the tables, and the smell of beer and whisky made a nauseating stench. In a passageway at the end of the room a dull red light glowed, extending a lewd and ghastly invitation to any and all to mount the obviously worn and rickety stairway. I wondered if one's blood pressure was supposed to rise in these surroundings. If so, there was something wrong with me, because mine did not. I found the environment ludicrous and sickening. So did many Hurley citizens, and, consequently, many attempts have been made to abolish such establishments.

We went to the end of the bar, where we each were served a glass of wine. I expected to see seductive women bewitching drunken but pathetically solitary lumberjacks. Then I remembered that this was Sunday night and that most of the jacks were on the road back to their respective camps. With no customers to please, the "broads" were out of sight. But before we finished drinking our wine, an unescorted girl came in and leaned wearily against the bar. She ordered a drink and with a wave of her hand indifferently indicated that she would buy one for me, too. I do not think she saw Howard beside me. A flush of embarrassment colored the bartender's face as he explained that I was with an escort. The girl's pale lips twisted into a grimace. Without trying to hide the contempt she apparently felt for me and my smug, respectable security, she lifted her glass and drank all of its contents without a moment's hesitation. Not even glancing in our direction, the girl dropped a coin on the bar, turned abruptly, and defiantly passed through the swinging doors to face an unfriendly, sordid street and a bitter-cold night. Her generous nature had manifested itself, and the rebuke must have cut her like a knife.

With pity I wondered what disillusionment had ever induced her to roam the then avaricious Silver Street in Hurley. Unlike the gay "hoop-shop" girls I saw later, who dressed garishly, wore gaudy jewelry, and used heavy makeup, she was plainly dressed and seemed extremely dejected and bewildered. This one pathetic woman slumped at the bar, drinking to numb her senses and make easy the utter destruction to which she had committed herself, for some unknown tragic reason, gave me enough of bawdyhouses for my lifetime.

After Howard made final arrangements for lumberjacks to be directed to his camp in the morning, we left. As we went through the swinging doors, a staggering jack came in. He nearly knocked us down and then began calling us the most shocking and abusive names I have ever heard. As we escaped his barrage of words and stepped to the street, Howard apologized for him. "Don't pay any attention to him. As drunk as he is, he is harmless except for his bad language. Some bartender probably fed him a quart of cheap whisky so a girl could rob him of a month's wages. He may have mistook you for the girl and me for the bartender."

"I'm not flattered, but I understand what you mean."

"If the jack had been sober and met us in this doorway, he would, I'm sure, have tipped his hat and held open the door." Howard spoke the truth, which was proved to me in the following years.

We went on to Doc's Place in the Burton House a block off Silver Street. I certainly was not—at first—impressed with the monstrous building. Undeniably dilapidated, showing the effects of rain, sun, wind, and snow, it loomed against the dark sky. It appeared extremely drab and ancient looking. The paint, what there was, was peeling; the stone foundation needed repairs; and I, thinking it might actually fall down on us, was afraid to walk under the sagging roof of the veranda. From the windows in the front of the building, shafts of pale-yellow light escaped from the sides of the blinds. Other scattered lights, showing the jagged outline of many dormers and cupolas, shone dimly from the second- and third-story windows. If it had not been for the friendly Coca-Cola sign above the entrance, I should not have gone in at all.

The Burton House appeared, in the blackness of night, to be the habitat of ghosts. Inside, however, Doc's Place was alive with gaiety, animation, and laughter. Over the chairs were a few mink wraps, casually cast off. The glint of diamonds blinded me for a moment, but I regained my composure and tried to brush off my schoolteacher look. A gray-haired, heavy-set gentleman using a cane approached us and asked Howard about his logging operations. Howard took my arm as he introduced us. We exchanged greetings. Doc's sincere smile made it apparent to me why so many people found his place in the Burton House the answer to their quest for congenial relaxation.

"Logging is going to pay this winter if the demand for veneer continues. I need sawyers," explained Howard. "I'm looking for another shacker or two. John Steffin said he would be living upstairs here until I had a shack repaired for him. Has he been around lately?"

Doc located the shacker for whom Howard was looking, and while Howard made arrangements with John for his transportation to camp, Doc found a place for me at the bar. When Howard joined us, Doc returned to his game of cards in the adjoining room. A short, partly bald, grayhaired man, probably about sixty years old, came to serve us a drink. He, too, knew Howard; and I met him—Frankie. I ordered a sloe gin fizz.

"Never heard of it," he sputtered. Under his breath I heard him stutter what sounded like, "D-damm f-fool womin's drinks. S-sloe gin fizz, bah."

"What have you?" I asked, with suspicions that he would not take kindly to mixing a fancy drink.

"Bourbon and Scotch." he muttered.

With shelves of many kinds of liquors in queer-shaped bottles on the mahogany back-bar before me, I drank a Scotch and soda and liked it. Frankie won the first round, which victory made us forever friends. When I knew him better, he came to serve me any queer concoction I chose to order—only mildly protesting.

We ordered sizzling steaks. Above the chatter, laughing, and the music of the juke box, I heard a rhythmic thud on the ceiling. Just when I thought I was imagining the noise, I could hear it again. Curious, I asked Frankie if he knew what caused the noise.

"Oh, that's S-Swan putting some lumberjacks to bed," he explained. "He's the clerk upstairs. It's his wooden legs that make all the noise." A draft of cold air swished my skirt. A broad-shouldered, red-faced lumberjack in a checked shirt and stagged pants appeared in the doorway. "That's S-Swan now, after his nightly bottle of beer." Frankie reached for the beer, loosened the cap, and handed it over the bar to Swan. Expressionless, Swan stood apart from the others at the bar and drank the beer slowly. Then he left, letting in another draft of cold air. In a short time I heard him thumping around upstairs on his wooden stumps.

I heard this strange noise many times on different nights before I asked Frankie, "How did Swan lose his legs?"

When there was a lull in business one night at the bar, Frankie told me Swan's story, which has become a legend in Hurley logging history commemorating a lumberjack's endurance. At first I could not understand Frankie because he stutters. I listened closely, and this is the story he told me.

Once Swan was a powerful man—one of the best Swede sawyers of his time. But he drank a great deal, like all lumberjacks. In a half-drunken state, he started back to camp in below freezing weather. There was no moon to light his way, which made tramping through deep snow difficult in his condition. He became lost, gave up the struggle to reach camp, and spent the night in a drift of snow. When he was found, his hands and legs were frozen. As a result, his fingers on both hands and his legs below the knees had to be amputated. Swan spent a long time in the hospital. To pay his expenses, incurred as a result of his lengthy hospitalization, a contribution was solicited from the generous lumberjacks and Swan's friends. There were no compensation laws in those days, so when a lumberjack or a miner was injured he had to pay his own doctor and hospital expenses. Kindhearted lumberjacks paid the bills of many unfortunate jacks who were accidentally injured or killed in the woods. To show his appreciation for what they did for him,

Swan acted as room clerk for the lumberjack quarters upstairs and put a great many of them to bed when they came in drunk.

Subsequent visits to the Burton House revealed to me that, while the lumberjacks occupied the upper part of the hotel, they never thought of frequenting Doc's Place. When by mistake a jack came in, he realized that he had strayed beyond his chosen environment and backed out immediately, obviously embarrassed.

On my first visit to Doc's, when it was not so noisy in the barroom, we could hear the roll of dice and the whir-r of the roulette wheel in the adjoining gambling room. After Howard and I had eaten our sizzling steaks, we decided to join the roulette players. Frankie cautioned me, "Don't throw away your money, Mrs. Anderson." Then, when I insisted on being the spendthrift for which I have the reputation, he advised me concerning which numbers were sure winners. "Twenty-four and thirty-six are the numbers that win."

To our left as we entered the card room, Pete, the blackjack dealer, sat behind a long, green, felt-topped table. His eyes, shaded by a green visor, glanced casually at us and then attentively back to the cards he was dealing. Before him were stacks of silver dollars. Silently and expectantly, the players brushed the felt with a card when they wanted a "hit." Emotionless, Pete paid the winners, while the losers reluctantly paid their losses. (It was Pete who eventually taught me how to play blackjack with the lumberjacks who came to the office at camp for cigarettes and candy bars. During the hours when Pete did not deal cards, he enjoyed fishing the Presque Isle River at the edge of our timber and hunting in the cedar swamp for deer.)

Along the south side of the room was a large round table at which a group of men were playing a game of poker. Nearby stood the table for shooting dice. The roulette wheel at the end of the room seemed by far the most popular spot in the place. One must know how to play blackjack and dice to win, but playing the wheel and winning, for me, is merely a motion of taking chips from a diminishing or increasing pile in front of one and scattering them about in a crazy pattern on the geometric design of the roulette table.

To make a science of playing roulette spoils the fun, I think; although I like to play seven, thirteen, and twenty-one. My lucky number the night of my first bet on the ivory ball was thirteen. In the glow of the low-hanging, wide-shaded lamp, I became hypnotized by the alternate quiet suspense, the noisy elation, or complete dejection of the players. While I lost what my first luck had won for me, Howard talked with friends. I knew by his animation that the subject must be logging.

When I joined Howard again, he told me he had just sold his hardwood veneer to Mr. McIvar, president of Penokee Veneer, at the highest price he had ever before been able to get.

"I'm glad we came here tonight." The excitement in his voice proved it. "I've been saved a trip to Mellen to sell my veneer. I've filled the bunkhouse with sawyers again, and another shacker is ready to go to camp on Monday."

During the years that followed, Howard and I combined business with pleasure and went to Doc's in the Burton House. This rambling hotel remained for a long time the landmark of the Upper Peninsula's pioneer days. It retained its decaying exterior dignity and its reputation for being fireproof until flames actually consumed it one Sunday morning several years ago. Hurley is not the same without Doc's Place in the historic Burton House.

After the Burton House burned, we found several nice places to go, where the sea food, steaks, and chicken were excellent, and the juke-box music remained turned down low. But my favorite place is a lodge on the shore of a small inland lake lined with white birch trees. A huge plate-glass window at the end of the room allows us to enjoy a delicious meal before a fire crackling in a tremendous stone fireplace and, at the same time, to view a north woods winter scene in all the beauty of moonlight. The atmosphere of the lodge is quite a change from the environment of Doc's Place. We have enjoyed both.

As we drove from Hurley back to Tula Camp one starlit Sunday night, we saw thin trails of smoke curl from chimneys to form elongated funnels extending up into the midnight sky until they became nothing. Naked branches of tall trees along the road cast slim shadows on the sparkling snow. For a moment a smoky blue-gray cloud passed between us and the moon, changing the lighting more smoothly than an electrician backstage could manage by using countless switches. With the same ease, full moonlight switched on again as the moon appeared from behind the cloud. The night became an ice-blue dream. We stopped before the office.

Tex came out to greet us with glad yelps while we tried to make our way quietly to the office. Howard tugged at the heavy door. A car came noisily into camp and swung into the parking place beside the bunkhouse. Its loud motor stopped. All was quiet. Not even Tex made a sound. Men came from the car toward the bunkhouse. One or two shadowy figures went in, and then long-legged Laggin' emerged from the car. Into the stillness of this moonlit night he let forth his strange wild yell. That sudden man-made howl must have awakened every jack in the bunkhouse. I imagine all of them were awake in time to see Laggin' come sliding down the aisle between the bunks on the seat of his pants.

Laggin' told us in the morning that he had slipped on the ice-covered sill. We saw him fling his hands high, trying to regain his balance. Then his feet went out straight in front of him, and he disappeared, sliding on his seat through the bunkhouse doorway. Roars of laughter and men's kidding voices followed. Then quietness and brilliant moonlight settled over the camp for the night.

Sawdust on Snow

Deep snow drifted around Watermelon's shacks that second winter at Tula Camp. High banks piled up along the road through the field and on our woods roads. As the snow in the timber deepened and drifted, the sawyers left the stumps higher. I expected my beauteous rose-breasted birds to return, and watched expectantly for their return, but if they did, I never saw them. I have learned since that it is not the habit of flocks of finches to winter repeatedly in the same conifer woods.

Once I came upon an olive-yellow bird of friendly habits. She perched on a low cedar branch and preened her lovely feathers. I watched her closely and several times came within six feet of her, but she disregarded my presence entirely. Low and sweet, her warble sounded from the thick hemlock boughs. A clear repeated whistle from a distant hemlock interrupted her sweet warbling. I am sure now these were a pair of pine grosbeaks. Had it occurred to me then, I would have attempted to see the male, as he is reputed to be a striking bird of brilliant rose coloring, not unlike the purple finch but much larger.

For me there had been no vacation from bookkeeping during the summer. And since October, when the crew reached well over fifty men, the reports had become even longer and more tiresome to compile. Fred Brandt lent me his adding machine, which I welcomed heartily, but I often wished it could multiply, subtract, divide, write checks, and lick stamps as well.

Finally I began the knitting which gives away the precious secret of every woman who is going to have a baby. All during the winter I became gradually more clumsy and spent less time in the woods, partly because the roads were too slippery and partly because my presence seemed to embarrass the lumberjacks. Even the most friendly, Emil and Ole, averted their gaze and pretended not to notice me. I know they were concerned about me, though, because when I slipped and fell in front of the cookshanty as I was going to breakfast one morning, six jacks appeared in a split second from half a mile down the road to help me to my feet. Falling embarrassed me, as I am and always have been slender and agile. Howard came running to help me, too. By the time he arrived, we were all laughing about my awkwardness.

On sunny afternoons Tex and I took walks along an abandoned woods road near camp. Howard thoughtfully kept it sanded so that I would have sure footing. I enjoyed these walks, but most of the time I stayed in the warmth of the office and hooked rugs, knitted, sketched, read, or worked at keeping the men's accounts.

A disappointment which I shall always remember was the loss of the toes of a pair of angora booties—the furriest, tiniest booties imaginable, which I had lovingly knit. When they were finished, even big burly Laggin' stopped in the office and howled with delight when he saw them. He slipped them on his thumbs and jokingly said, "They'd sure make nice warm thumbs for a pair of mitts."

Howard and Ed were of the opinion that they were too small. "Are ya sure babies come in that small of a size, missus," Ed said, doubting their usefulness.

"Our Jack has to start out with bigger feet than that or he won't get any place, Julie."

When the booties were finished, I left them on the counter in the office. Sometime during the night a mother mouse with intentions of building a nest for her babies came to find some fluff to keep them warm. Silently she chewed the toes out of both booties. Had she been able to do so in one night, I think she would have carried them away entirely.

In the morning, when I discovered the damage to my precious booties, I vowed to wage a war on the mice. I set traps all over the office. Never did I forgive that mother mouse for chewing up the booties, but they have been mended and worn by all three of our boys. No one has ever noticed the mends.

The mice did a lot of scampering around at night. They nibbled the leaves of every plant I took to camp, until I finally gave up trying to keep plants at camp. They carried away peanuts and chewed at the candy bars under the counter. Industrious mice they were, for in the morning only empty sacks remained, looking like perfect counter displays minus the contents. Along the low rafters the mice dropped many peanuts, revealing, like Boy Scout markings, the paths they had taken to their nests.

I never was able to track them down until one day Howard wanted some parts he thought were in the big box we used as our davenport. I began lifting the things out one at a time—boxes of bolts, spark plugs, newspapers, and other odds and ends. In one dark corner I came upon what appeared to be the part Howard wanted wrapped in cotton. I closed my hand over it. Excited squeaking came from the wad of cotton. Its contents wiggled and squirmed. In the light, what I had in my hand proved to be a nest of tiny pink mice. Yes, there *was* a bit of angora fuzz—and strips of paper, and tiny pieces of cotton. Kicking the door wide open, I threw the nest of squirming, squeaking pink mice out into the snowbank. Heartless of me, no doubt, but what else was I to do?

Shortly after that the adding machine ceased to function. Howard worked over it, oiled it, unscrewed screws, screwed some screws tighter, and, at last, sent it to a repairman to have him look at it. When the bill came for fixing the adding machine, the item we were charged for read: Extraction of mice nests.

Christmas payday time in 1940 was a repetition of other Christmas paydays: big checks, bonuses, laughter, and kidding. Snow fell all day in big flakes that piled up quickly and covered the crusty snowbanks. Christmas day would be a picture postcard day. The jacks would be gone only a few days because of the great demand for lumber products.

A wooden ash tray made from layers of veneer treated with plastic arrived in our Christmas mail. It was sent from Penokee Veneer Company as Christmas greetings from Mr. and Mrs. McIver. This verse was attached:

> This is a birch tree all cut down,
> Treated with plastic, golden brown,
> That made the stock
> To carve the "prop"
> That flies the plane
> That drops the bombs
> On HITLER!
> MERRY CHRISTMAS.

—Simple poetry but with a meaning it was going to take years for us to comprehend.

With thoughts of bombs being dropped on helpless people in England and Germany to sate the passions of a maniac, we tried to enjoy our simple Christmas celebration in our apartment in Wakefield. Shortly we were back at camp, cutting those same birch trees that went to Penokee Veneer and eventually became the props of planes flying over and dropping bombs on Germany.

In January of 1941, the sawyers' strips were along the far section line, nearly a mile from camp. Howard decided that because the distance from camp to the timber was so great, it would be better for the men to eat their noon meal in the woods. Thus, time would be saved and the men would not be tired from the two-mile hike to and from the cookshanty. Eating in the timber did not offer the comforts that coming to the cookshanty did, but the new cook made their lunch as

"Dinnering Out"

appetizing and complete as he could. In one box he put the plates, cups, and tableware; in another he sent kettles of hot soup, pots of coffee, hot dishes of macaroni or spaghetti, plenty of roast meat, mashed potatoes, and gravy. In still another box he packed cake, cookies, and sometimes spicy pumpkin or apple pies. The cookee helped the cook load everything into the new pickup truck or on a dray pulled by one horse. This dray was called a "swing-dingle" by the old-timers, because it resembled the dray used in pine logging days when jacks "dinnered-out." A regular swing-dingle, I might add, had a shelf that swung out to serve as a table on which to put the dishes and food.

The lunch arrived in the woods steaming hot. How it warmed our lumberjacks who gathered around the kettles and bowls and helped themselves! No one dawdled over the meals served from the swing-dingle. The hot food was quickly consumed before it had time to get cold. As soon as their plates were empty and they had smoked a cigarette, the jacks all went back to their jobs. Working kept the men in the woods warmer than hanging around a small fire in the clearing.

"Ya sure know how to git them men workin'," Ed jokingly accused Howard.

Later that winter Howard built a lunch shack on skids, which could be moved around wherever it was needed. It was not large enough to accommodate all of the men at one time, but it kept out the raw wind, and the food stayed hot longer. No one thought to complain. Dinnering out was a situation to be endured, as such procedure for loggers was not unusual. It had been a daily occurrence every winter for years in logging camps of Lower Michigan when pine stands still covered that region.

After they had been in the timber all day, Tula Camp looked mighty good to the jacks coming back in the late afternoon as the shadows lengthened and the sunlight faded. Laggin' howled as he swung into the warmth of the bunkhouse. Ed and Howard stood around the stove in the office; sometimes they would be joined by Mr. Rogers, Penokee Veneer's scaler, or truck drivers or jacks wanting snoose and candy. Thus they shook off the all-day chill which had settled upon them and given them frosty eyebrows, red noses, and cold feet and hands. They all gathered around the stove, warming their fronts and backs alternately, and rubbing their cold hands together.

"It's a good life, here. That it is, Missus," Mr. Rogers always reminded me as he soaked in the warmth of our barrel stove.

Looking back, I find those years pleasant to remember, especially the frosty kisses and big hug Howard never failed to greet me with when he came in from the woods at any time.

"I'm certainly lucky you don't mind staying in this isolated camp, Julie," he would say. "What did you do alone all day?" And then he would flop down to rest before the supper bell. And I would proceed to tell him of the birds I had seen on my walks, or about the book I had finished reading. Sometimes I would show him my sketches for his criticism.

In February, Howard put a gang of sawyers in on the strip where the big pine stood—the pine I had visited on a number of occasions in summer, winter, fall, and spring. "I put Carl Spray and Jack Coleman in on that strip. They'll get the big pine down without any trouble," Howard said.

That next Saturday after supper, Carl and Jack burst into the office. They wore smiles from ear to ear. "Ya know that big pine? There's three pine trees there, but one's too big for us ta saw with the blade we got. Ya won't believe it, but we got it notched and it jist can't be felled 'till ya get an old pine saw fer us," Carl said.

"Aw, heck, that pine tree ain't that big, is it?" asked Steve the teamster.

"Sure as you're a standin' there, it's that big. And we ain't a gonna fell it till we gets a saw that'll do the job proper," Jack said, substantiating Carl's claims about the big pine.

"I'll bet ya a five spot it ain't thet big—nowheres near it," said Steve, who was always ready to make an easy dollar or two.

"Cum on out there and we'll prove to ya how big thet tree really is. I hain't seen anything like it since I was over by Merriweather cuttin' for Humphrey."

They left the office to walk out to the pine and settle the bet. I held their money and listened to them argue until they disappeared around the curve beyond the barn. When they came back to the office, Howard said, "Well, boys. How big *is* that pine? Big enough so I'll have to borrow a saw to cut it with? I haven't got a pine saw left in my rack."

Steve threw his cap on the floor and swore. "Damned if it ain't the biggest pine I've run across since I left Manistee. Give them guys my dough, Missus."

Howard drove over to Brandt's Landing and borrowed Fred's big pine crosscut saw. Carl and Jack spent Sunday morning getting it sharpened to cut the giant. When they were ready to fell it, half the crew, Ed, Howard, and I went along to watch. We waded through deep snow—Howard pulled me out of snow-covered brush piles when I clumsily got caught—to where the tall pine stood with—I gasped—a huge notch cut in its broad base. I bent my head back and looked up and up to the top of the pine skyscraper of our timber, which had not changed since the romantic day Howard and I found it. A strong cold wind blew around me. Above the men's excited voices and Tex's barking, I listened for my great pine tree's haunting plaintive song. Its familiar melody came to me in minor tones, and I strained to hear its last fading chord.

Carl swung the saw from his shoulder. It glistened in the bright sunlight. On either side of the pine behind the notch, Carl and Jack found firm footing and went quickly about bringing this giant to the ground. I doubt they heard the song of the pine. They drew the long crosscut saw back and forth through the unnotched side of the tree. Yellow sawdust arched out onto the white snow. Men's purple shadows stretched out upon the glistening drifts. Pine smell stung my nose, but I breathed deeply of the frosty air and pungent odor. Tex stood patiently behind Carl and Jack. Other sawyers stood some distance away, talking low and motioning in the direction they expected the tree to topple. Howard pointed the movie camera to the top of the pine to record its magnificence and its end.

Carl, with the face of his ax, pounded hardwood wedges into the cut behind the saw. The sound rang out musically into the clear cold air, and a woodpecker answered with a "twack, twack, twack." The sawyers resumed their cutting and pulled the blade back and forth through the centuries-old landmark. Now Tex crouched into position for his dash to the quivering branches. The watching men became silent–listening and awed.

Carl straightened up and looked in the direction in which the giant would fall. He scanned the sky where the pine's tiptop branches, brushing the clouds, swayed slightly. "I think it's about ready to topple," he called to us. I stepped back. Howard aimed the camera. Tex's ears went up, and Carl and Jack pulled the limber blade through for the last cut. A crack sounded! The sawyers pulled their blade out and backed quickly away.

"TIM-BER-R-R!" sounded through the forest in a mighty way that made the hardwoods tremble. Carl had put his heart into that warning.

The saw cut widened, and another loud crack cut the air! The saw cut rapidly widened, while the cracking noise became louder and louder! The giant leaned, groaned, and came crashing through the hardwoods. The immense tree rolled away from its stump and dropped heavily into the snow and brush in a shower of broken branches and huge pine cones. Tex leaped and bounded off to shake the branches and growl. Men gathered around the stump, talking of this big pine and of other giants they had once felled. Across the smooth yellow butt someone chalked in big blue numbers the giant's diameter –5′ 4″!

From a distance I watched, completely stunned at the tremendous vacant space in the forest our big pine left as it lay felled.

Jack and Carl stayed to limb and saw the pine into log lengths, while the rest of us went back to camp.

"It's too bad to cut a tree like that, isn't it Howard?"

"Oh, I don't know, Julie. That tree might have stood for another hundred years and then maybe another hundred; who knows. But, remember this, too, tomorrow it might have been struck down by a bolt of lightning and become splinters and broken limbs and branches. Carl did a good job of bringing it down. Maybe we can save some of this pine to put with that other pine we're saving. To make you feel better about cutting your big prince here, we'll put some white pine trees in our yard that won't ever have to be cut down. How's that for a promise?"

Howard kept his promise. We have today three beautiful pine trees in our yard, which we tramped the woods to choose and which we brought in on a truck to our lot in Wakefield. Their minor songs are not so throaty, but they are just as beautiful and haunting.

Late that afternoon Carl and Jack came into the office, hung the big pine crosscut saw in the rack, and reported to Howard that the pine was ready to drag out of the woods to the skidway. "Got it all cut," said Carl. "There's logs enough in that one tree ta make a load that'll break all the records this year."

During the night a heavy snow fell. When Charlie and Steve went out in the morning with the tractor and the team to skid the pine logs, a thick blanket of fresh snow covered everything. Snow lay heavy on the hemlock boughs and on the leaning rampikes and downed trees. The moss on the birch and maple trees held patches of fresh snow. Fungus and stumps supported deep chunks of overhanging snow, too.

After lunch the sun broke through the gray skies. I drove out with Howard to watch Melvin and Red load the pine logs. Because they were so big around, Howard found it necessary to roll them onto the trucks with chains instead of using the regular grab hooks. Steve fastened the evener behind the gray team onto the jammer cable, and slowly the smooth pine logs rolled over the skidway, slid up the skids, and dropped into place on the trailers. Five choice logs made the first load a sky-high load. Laggin' pulled the binders tight around the logs and checked the rigging. Joe swung into the cab and carefully pulled away from the jammer, while Steve called out, "Bet a five spot that's the biggest load ya git ta haul this winter."

No one took the bet because we all thought the same thing Steve did. When Joe came back from the mill, he proudly waved the scale sheet totaling 3,960 board feet. This load gave him the distinction of hauling the biggest load of choice white pine cut that winter in Gogebic county. The smile on his face as he swung from the truck to load his second load gave evidence that he had a record. We were proud, too.

We have a complete film of the felling, skidding, and loading of the big pine. It thrills me every time I see it, and we never miss an opportunity to show our friends our "Big Pine" film. The memory of the big pine, however, as it stood in the forest, without a huge notch marring its broad trunk, thrills me more.

Not many weeks after the big pine went to the mill, the weather became warmer, the snow began to melt, and the men worked extra hard to get all the logs cut and skidded out of the woods before breakup. Crows came back to the scraggly pine behind the barn. Roads became soft and muddy. One day the last loads of logs rolled by the office. The crew buttoned 'er up on the twelfth day of March that spring. Tex and Axel, who had stayed to take care of the camp the year before, remained behind again this spring. The jacks all left, Howard and I went back to our apartment in Wakefield, where I waited, not at all patiently, for what I hoped would be our daughter. I regretted not being able to go with Howard to the woods when summer logging began and to Brandt's Landing as I had done the summer before. Instead, I counted every load of logs that clangored by on the railroad spur past our kitchen window.

Tightening the Binders on a Pine Load

Part Five

Curves in Our Trail

On a May morning in 1941, the clatter of hospital wash pans being banged around and filled with a sudden rush of water woke me. I am in the wrong place, I thought, for I missed the familiar sounds of the cook's triangle or my gentle-ringing alarm clock to awaken me. But one look around the hospital room and down at the flatness of the covers assured me there was no mistake. This was the private room at Grand View Hospital we had arranged for months ago; and if my suspicions were right, Howard and I were now a family of three. I thought of Howard and remembered his gentle concern on the drive to the hospital and the look of anguish in his eyes when the nurse wheeled me away to the delivery room.

Contrary to the rules of the hospital, Howard appeared in my doorway only minutes after I had eaten my breakfast and had had my first glimpse of our blanket-wrapped baby boy. Howard brought with him a mass of purple violets planted in a white bowl.

"Where on earth did you find these beauties? They're lovely, Howard. These deep blue petals are exactly the color of our baby's eyes."

"I couldn't sleep, so I went out to camp early. I dug the violets from the woods along the Presque Isle River where it cuts the far side of the section. So the baby has blue eyes—like yours? I haven't seen him yet. Doc says he's O. K. Prettiest baby he ever spanked."

—Some days later a letter came to us which I prize very

much. Howard, proud of our new son, Christian Howard, made a note of his birth on a remittance to a wholesale dealer in hardware and loggers' supplies in Ashland, Wisconsin. This is part of the letter we received:

"My dear Mrs. Anderson and Howard: Following up the note attached to your nice remittance this morning, will say that I immediately called a meeting of the Top Loader's Union, and by unanimous vote this young son and heir of yours was elected to its membership. The name and number of this order, of which I am president, is Crooked Steel Order No. 7-11. Good wishes, Will Garnich."

We brought Chris home to a nursery I had proudly planned for his and my convenience. I had transformed our small extra bedroom into a blue-and-white fantasy. Our guests would now have to sleep on the davenport to give way to our son.

During that summer Howard asked for an option on a section of timber north of Connorville, near Lake Superior on the west side of the Presque Isle River. The shackers were scattered over the section to cut the last of the short stuff at Tula Camp, and Howard settled down to logging the last of the hardwood. Caring for Chris absorbed my entire day. He grew fat and jolly, a truly beautiful baby, and one entire summer passed rapidly.

In September, Howard faced the problem of buying more timber. "Well, Julie, the last loads of logs will go out of Tula Camp tomorrow. There's some short stuff left to get out, but I guess tomorrow's hauling will clean up the hardwood and pine logs. My contract with Marathon is filled at last. I'm going out tomorrow to look that section I've got under option. It's lots rougher country than Tula country, but if there's enough hardwood on it, it'll be a good buy. I'll get Fred Brandt to cruise it with me. Wish you could go along."

"I've got to take Chris up to have his shots. I wish I could go, too. Maybe later when the trees turn color I'll get a chance to tramp along some afternoon."

Fred and Howard came home late the next day, well satisfied with the timber they found.

"You should see the beautiful pine on the back of that section, Julie. None of them are as big as the big pine at Tula Camp, but they sure are nice ones. There's not as much veneer as I would like to have. The hardwood's all good though and there's lots of hemlock and cedar."

And so Howard bought Section Thirteen.

He tore down Tula Camp buildings, all but the office. Trucks hauled the lumber and logging equipment to the new section. "Camp Seven," Howard called it. The office was jacked up on skids and, just as it stood, was hauled slowly into the new camp site. Water, piped from a clear cold spring, furnished water for the cookshanty and the horse trough. Howard laid out the roadstrips and sawyers cut that timber. Tractors and trucks roared through the timber, building roads over streams and through gullies. In a stand of cedar Howard had a space cleared for Arvid's shack. He located other shackers where they, too, could cut small stuff. Their shacks were hauled to Camp Seven in the same manner by which the men brought the office, on skids behind a truck.

The October days of 1941, bright with vibrant colors, lured me out to the new camp. That one day was sheer magic, warm and dreamy and still as a summer night. The spell of virgin timber, which has always enchanted me since the first hike Howard and I took, seemed stronger than ever before. That day and the joy in Tex's eyes when he saw me convinced me that we three, Chris, Howard, and I, could live just as cozily in the office as we had before. Howard agreed, and Ed the foreman decided to stay in the bunkhouse. While the weather remained nice, we put compo-board on the office walls to make it warmer. We cut an outside doorway into what was our bedroom, and made the old office our living quarters. Our old bedroom became the office. With Chris watching me from his buggy, I painted the plank floors gray. Howard replaced the putty around the windows. And best of all, he built a small addition where he installed a chemical toilet.

With fresh starched curtains at the windows, a new maple chair, an end table, radio, and lamp, and an oil-burning stove to replace the barrel stove (we thought Chris might get burned easily on the barrel stove), we transformed the office into a very pleasant place. For Chris we found a crib that was high off the floor for warmth and big enough for him to play in. Tex deserted our bed and chose to sleep under Chris's crib, which delighted Chris and pleased me, too.

Because of the shortage of labor, Howard had to get along with fewer men. George and Irene Holmes came to take care of the cookshanty. As usual, by the end of October sawyers had begun to fell the timber, and from the woods came the sounds of crashing trees and the roar of skidding tractors.

The weather stayed warm. Heavy snow did not fall until late that year. Trucks came out of the woods with big loads, making as many as twelve trips a day. We discontinued sorting out the veneer; there was so little it did not pay. The hardwood now went to Milwaukee, where the Quirk Company converted it into a vital war product. The hemlock still was piled on Brandt's Landing, eventually to become tannic acid and paper pulp.

George and Irene, a friendly couple, kept the cookshanty immaculate. I enjoyed having feminine company, after years of being the only woman at camp. While Chris slept after I had bathed and fed him, I went to visit with George and Irene and have a fresh roll and coffee. Howard usually came in, too, for coffee. But he was much too busy to stay long. In the words of Irene, "He comes in. Pours himself a cup of *hot* coffee. Fills the cup with *cold* water. Takes one swallow, and then, by gosh, he goes and dumps it all in the sink."

During those short winter afternoons, while Irene and I sipped our coffee, George shaped a batch of rolls or trimmed the steaks. While he worked he told of other days when he cooked in logging camps along the lake. The Schroeder Lumber Company for whom George had cooked was one of the big lumbering concerns that logged heavily in the Upper Peninsula region and used the icy waters of Lake Superior to float their logs to the mills.

"Thousands of feet of logs chained into rafts were towed by tugs to Ashland. On one night, thick with fog," George told us, "a blustery stubborn captain of the tug *Butterfield* met up with a stubborn captain of an ore boat. They approached each other head-on and neither would give way to the other. Both cussed and stormed and held to their courses.

" 'If ya don't get your damn toothpicks outta my way I'll ram 'em,' the ore-boat captain shouted to the captain of the *Butterfield.*

" 'Ram 'em and be damned,' was the answer.

" 'Ram 'em? I will that.'

"And ram 'em the captain of the ore boat did. Soon he felt his ore boat being towed miles off its course. Held fast by the raft of logs, he was carried six miles before the *Butterfield's* captain relented enough to break the raft and let the ore boat free.

"Storms on Lake Superior are far worse than on any other of the Great Lakes," George said. "One captain of a tug towing logs from Superior's north shore, when Schroeder cut timber there, logged his course during a bad storm. Five times it showed he was directly off Isle Royale in five hours. The gale—the worst ever recorded in the Lake's history—had kept blowing him back. But the tug finally reached the breakwater in harbor at Ashland with the log raft intact."

It was George who showed me how to make oyster stew without having it become so curdled it looked like cottage cheese. Many times when Howard went to town for the mail or tractor parts, he brought back fresh milk and oysters. If there were no repairs to be made on the trucks and tractors after the jacks had come in for their cigarettes and candy, we tucked Chris into his crib for the night and went to visit Irene and George. Irene and I would let George and Howard win a game or two of cribbage. Then we would "up and skunk 'em," as Irene said.

While we played our last game, the oysters, sprinkled with salt, pepper, celery salt, and paprika, and dotted with butter, would be curling their edges in one pan at the back of the cookstove; and the milk, in another pan, would be slowly heating to just the right temperature. Shortly before the time to turn out the lights, George would pour the milk over the oysters and ladle out into warm bowls the best oyster stew imaginable. His final flourish, celery salt, made the perfect seasoning, we told him, as we ate every drop of stew.

With the wind howling and blowing the snow against the windowpanes and rattling the stovepipe chimney, we felt cozy and secure around the glowing barrel stove in the cookshanty on those long winter nights.

The days were pleasant and leisurely enough for Chris and me. For Howard they were long hours of hard work. The need for logs became the most urgent in history, and lumberjacks and loggers did all they could to supply the demand. I did not go to the woods often, but when the men came to the office for van I could see that they were restless. They talked a great deal about the war, which was so close but not yet an actuality for them. I sensed that they felt they were marking time—for what and how long?

Every Sunday we drove to our apartment in town so I could wash Chris's things. When we returned to camp late in the evening, I would hang the entire four dozen diapers in the office to dry, using all the available space there was by crisscrossing the lines. Jacks who came in for a supply of tobacco or candy learned to enter the office with caution on Sunday nights or get wetly slapped in the face with diapers.

One Sunday night while I was pinning up diapers in the office, we were suddenly jarred into the actualities of war. Breaking into our favorite program of remembrance music came jumbled and excited interruptions. Amazed, we refused to believe these accounts of what had happened earlier at Pearl Harbor that day—December seventh!

"War! War with Japan!" Howard shouted to me.

"Oh no! Are you sure, Howard?"

"Yes. There's been a surprise attack at Pearl Harbor!"

This news had the same effect on our small home in a clearing in the timber as it had in millions of other homes. We stayed up half the night listening for reports and talking.

The young jacks and Wisconsin farmers gradually left camp, one by one, to join the Army, the Navy, or the Marines. Deferred? "Hell, no!" they said, "we want to fight this battle where we can see it." So they came to the office for their last checks. Laggin' the swamper and Joe the truck driver left together. Their good-byes were awkward but sincere. Laggin' admonished me, "Take good care of Chris, and be sure not to take any of them there wooden nickels, Missus." We kidded Joe about putting his shiny new truck in moth balls, and down the road they went— around curves, over hills, across an ocean to Africa, to Sicily, and then through the blood-drenched hills of Italy to Rome and on to Berlin. They went down the timber-lined road from Camp Seven with only one thought, I am sure—to get the battle won and then come back to the tall trees.

Marvin, Red, Ray, and Hayward, when he became old enough, and many others all followed Laggin' and Joe down the road. Steve the teamster crowded his age and won a five-dollar bet. I wished he had lost that bet, because I never saw him again after the day he came whistling up to the office. He brought a cellophane-wrapped teddy bear he had won on a punch board in Hurley for Chris to coo over. Strapped to his back bulged his worn and stained packsack. With his hat in one hand, Steve pumped my arm up and down in that last farewell. He held Chris for a while and got his shirt wet for the trouble. Then he said, "Good-bye, Missus. Sure been real nice to know ya. See ya again after I get this here scrap settled." Steve went down the road like the others, but he followed it only around the curves, over the hills, and across the ocean to Africa. There, as we were in time to learn from Joe, this earthly trail ended for Steve.

So only a few old jacks and Howard were left at Camp Seven. We stuck to logging. Log prices went up, and controls began. Labor costs increased. Food prices went up, and rationing began. Our trucks and tractors broke down. Parts were not available. Priorities delayed repairs. Men who could work in the woods were few. And all the while, other men were fighting and getting pushed back and killed. They were, like Joe, defending ground they had never seen before and taking, time after time, some hill on which every stick and stone and scraggly tree was as foreign as the mire that sucked at their legs and the rain that never stopped coming down. They were, like Hayward (as we learned from his frequent letters and messages from friends), riding astraddle a Packard motor in the belly of a mosquito boat, sweating, and delivering their Sunday punch; and then streaking for cover in a bay—or sputtering home shot to pieces. They were, like Red, listening to the eerie silence of the jungle at night, then the sudden squawks of parrots and the banshee cries of crazy men. They were, like Laggin', riding in armored tanks with infernos surrounding them.

We did what we could. Howard worked nearly night and day, but it did not seem to be enough. Log prices climbed still higher. Labor made more demands. The war became grim and rugged. Thus began the years which seemed like centuries to those of us who waited and labored at home. They must have seemed like eons to the men who fought.

Howard and the truck drivers made repairs on equipment at night so that logs could roll out of the woods with regularity all day long. With fewer sawyers to work in the woods and with demands for logs increasing every week, Howard attempted to meet the situation by buying two power saws. Don and Einar, both good crosscut sawyers, recognized the power saw as a machine built for increased production of logs, and were the first to cut profitably for Howard with it. It cut faster, made a "helluva" noise, and was more cumbersome to carry through the timber than a crosscut saw. It took Don and Einar weeks to become accustomed to using it. There were times when, in order not to be injured by falling limbs, they had to drop the saw and get away in a hurry. Dropping the machine resulted in broken handles and tail stocks.

Howard reinforced the tail stock, changed the position of the clutch control, and put rubber between the handles and the motor to reduce the vibration—all this to promote faster and more felling of trees. In order to decrease pinching the chain saw, the sawyers began using thicker wedges. They learned to be more alert for falling timber, because the increased noise reduced warning sounds that preceded falling branches.

Gradually, after months of trials, experiments, and changes in makes of machines and methods of felling, production per man began to increase. Howard said the biggest saving, however, was in timber stumpage. Sawyers now could cut trees much closer to the ground, leaving less stump to rot in the woods. They also cut questionable trees that crosscut gangs might have left. In a matter of a few minutes the power saws felled trees it would have taken thirty minutes to fell with a crosscut saw.

At Camp Seven we were not awakened on Sunday mornings with Watermelon's Italian songs and garbage collecting. Instead, the bears raided the garbage cans and carried away everything George threw out. One bear disappeared with two suits of underwear and tore to shreds a red plaid shirt Baldy hung out to dry. When one black brute became so brave that he climbed a tree near the cookshanty, let himself down onto the roof, and tore off the tar paper on the meathouse roof, Howard called the State Police.

A posse of blue-clad, gun-toting police and curious lumberjacks sat around the cookshanty early the next evening drinking coffee. When they heard the brush crack outside, the troopers stealthily crept out the front and back doors, con-

Spotting "Widow Makers"

fronted the surprised bruin, and "shot him dead" with no trouble at all. Howard skinned him, buried the carcass with the bulldozer, and nailed his thick black hide, with head and paws attached, to the barn door. All winter until the weather turned warm, that black fur hung there, and we were not bothered again by raiding bears. I am inclined to think that what we supposed to be *bears* was really only one—the one that got shot.

Howard began employing more and more married men, who drove to Camp Seven from Thomaston or Wakefield each morning and returned home at night. The old-time lumberjacks occupied only half of the bunkhouse. I missed the young fellows who had always joked and kidded one another and Howard and me. In the spring, as before, we went back to Wakefield to live in our apartment. When the breakup ended, the summer logging activity began again, and loads of logs came out of the woods covered with mud. Howard came home evenings exhausted from the heat, but he always left early the next morning to start the crew into the woods. Having Chris to take care of and a home to manage—washing, ironing, cleaning, and cooking to do and dishes to wash (things I never did at camp)—made these summer months fly by for me.

Suddenly another October arrived—the time of the year that I like best. Chris, Howard, and I again established ourselves in our comfortable office at Camp Seven. We drifted easily into the old way of living at camp. I kept the books, played with Chris, did some reading or sewing, and resumed my visits with George and Irene. Chris entertained himself with George's many pots and pans or with the tamest of the increasing number of cats who came and went out the back door of the cookshanty. Frequently I left Chris in Irene's care and hiked the woods roads as I had done at Tula Camp.

Logging was tough for Howard that winter. The rocky gullies and deep ravines at the back of Section Thirteen presented a much more difficult logging chance than we had encountered at Tula. The roads Howard built through this rugged country were more costly than he had estimated they should be. Then snow began to fall and never stopped falling, it seemed. Every inch of snow meant greater logging costs for us in loss of timber by the high stumps the sawyers were forced to leave and by the hours Howard and the tractor drivers spent keeping the roads open for hauling. Breakdowns of equipment caused additional expense and reduced our profits. Howard was forced to sell some of his equipment. He kept only one tractor and two log-hauling trucks. When breakup came in 1943, we possessed little except our remaining timber, our camp buildings, and depleted logging equipment. Such were the results of all the bad breaks we had experienced.

We moved into a rented house in Wakefield in the spring where our second son Hans was born. After that time we never went back to camp to live. To be established in Wakefield was more comfortable for me than for Howard. He found traveling back and forth every day to camp difficult and a waste of time. But that problem loggers have always contended with—and so did Howard. Of course, it was not as easy for him as living at camp had been; but in winter he found the comforting fire from our own hearth just as warming and cheering as any barrel stove, and in summer he decided that the cool sheets and our well-ventilated rooms were more restful than our hot, mosquito-infested office at the Cedar Camp had been.

As the demand for lumber continued to increase, the log market looked better. Howard took heart again as his dad had always done. He invested what money he could spare in more timber surrounding Section Thirteen. To get a new perspective on logging, we flew over our timber. Howard, who had always been interested in flying since his first plane ride in Chicago when he was thirteen, had earned his pilot's license the summer before. When flying conditions were good on Sundays, we flew over Camp Seven and the Presque Isle area where Tula Camp buildings had stood.

After several such trips, we concluded that cruising stands

of timber by airplane cannot take the place of looking new stands as the oldtime cruisers do. By making use of areial photography a general estimate of the predominating species on a large tract can be easily arrived at. But a logger who has to operate on a narrow margin of profit wants to know quite accurately the amounts of veneer, Number Two hardwood, pulpwood and mining timber and other short stuff a tract of timber will produce. This estimate can only be arrived at by tabulating an average area for its potential value of timber products—and, using this estimate as a basis, evaluating the entire section. Most important in a logger's estimate of timber value is the comparative number of unsound trees in a stand.

"Some people say aerial photography's replacing the old method of cruising timber, Julie. But no photograph will ever reveal to me what a good strong whack of an ax on a tree can tell a cruiser," said my logger husband one day, after looking over the snapshots he had taken from high up.

Howard built more roads through the timber he bought that summer. In the fall he hired four gangs of gypo sawyers sawing with crosscut saws, and three gangs who used power saws. Where the tractor drivers punched in skidways, the teamsters decked logs for winter hauling. These piles grew steadily all summer. When the winter hauling season began, a half-million feet of hardwood logs lay waiting to help fight our war.

Arvid and the other shackers built themselves shacks farther back in the timber on Section Thirteen. One shacker now used a one-man power saw to fell and buck his short pieces. Howard worked from five o'clock in the morning until late every night. Many times, when he found it necessary to make repairs on trucks and tractors, he ate supper with George and Irene and did not come home until long after dark. Several times during that winter's snowstorms, he stayed all night at camp to help plow the roads and clear the skidways. His face became thinner. I had to take tucks in his pants so that they would stay up on his slim hips. Again, as after his dad died, he began doing the work of two men; and gradually logging profits increased. With his old enthusiasm, he replaced the equipment he had sold. Logging had a zest to it again.

Two more years passed rapid and profitable. Our third son was born in May, 1945. The balmy summer evening we drove to the hospital I remember well. A week later I returned home to Chris, Hans, and Howard with our new baby in time to enjoy my late-blooming tulips, which I had planted the fall before. We decided to name the baby Hayward after my brother—stationed at that time at a Navy base on Samar Island. With two small boys and a baby to care for now, I hardly ever found time to go with Howard to the woods.

Howard was logging with all the available manpower he could hire. The crew of old-time jacks had become so small that George and Irene were no longer needed to keep the cook-shanty going. Those jacks who stayed in the bunkhouse cooked their own meals and kept their own fires.

Finally the end of our war in Germany came. For many of our friends, the armistice offered an excuse for noisy celebration for which there had been little occasion during war years. Howard and I just stayed home, thankful and glad that security was about to be restored to us. We prayed for quick victory in Japan. Howard logged harder than ever.

Then came the atom bomb! At last the big words, "Japan surrenders," telling of the end of the war in the Pacific area spread out before me as I unrolled our daily paper. My tears of thankfulness came so fast that the big words were all I could read.

Immediately Howard and I began to talk of the return of Hayward and the rest of our old crew. We had not heard from Laggin' recently, but we knew Joe was all right according to his last card. Hayward's last letter told us he was still at the Samar P. T. base. We had heard that Red was heading for Japan. Einar had been exempted because of his age; he had the job now running one of Howard's power saws. Marvin, Ray, Bob, Clarence, Laggin', Hayward, Red, all but the betting Frenchman, Steve the teamster, finally came home alive.

We thrilled most to see Hayward in his dress blues swing off the train in Ironwood. He did have fewer fingers than when he left for Samar, but his big boyish smile spread over his browned face in the same old infectious way. His eyes glistened at the joy of being back in our north woods. My eyes filled with tears at the happy fate that brought my brother safely home.

"Gee, am I glad to see you folks. Where's my namesake? How's loggin', Howard?" From that moment on things were for us all as if he had never been away.

A Cycle of Three

Our years together brought Howard and me much happiness and some despair; but the year of 1946 became the blackest we had yet known.

Early in January an injury to my spine of undetermined cause (quite likely suffered years before, the doctors thought, when I had slipped and fallen hard on the ice in front of the bunkhouse door at Tula Camp) began to cause me such extreme pain that I could no longer take care of our boys. Lifting the baby into his bath became agony instead of the pleasure it had been. Simple housekeeping tasks caused pain that left me hunched like a witch and miserably unhappy. The effect of my helplessness on the children and Howard prompted me to enter a clinic in search of relief.

We found a housekeeper to care for the children and drove four hundred miles to a clinic. After days spent in torturing positions for simple X-rays and preliminary examinations, a clinic surgeon told me it would be necessary to take an X-ray to determine the nature of the damage to my spine. For that X-ray I entered a hospital co-operating with the clinic. Drugged and blissfully without pain for the first time in several months, I submitted to a spinogram, the most excruciatingly painful of all X-rays from which to recover. Moving my eyes from the nurse to Howard's tired, worried face gave me the splitting headache I had been warned about. I lay motionless all night, thinking and hoping the X-ray would reveal a protruding disc that could be removed and give me sweet painlessness again.

In the morning, while Howard stood nearby, the doctor quietly told us, "There is no protruding disc. We think two vertebrae are crushed together and are pressing on the nerve that causes your pain. An operation would bring some relief —not a complete cure." These words completely crushed our hopes for my recovery from this crippling pain.

Howard sat with his blond head bowed in his thin hands. I stared at the ceiling.

Later I was released from the hospital for further examinations and consultations at the clinic. Alone there, one day, in a cubicle used for such purposes, sobbing with pain and blue with cold (two thin pieces of sheeting were all the nurse left me), I waited and waited.

After a time two doctors brusquely entered, directed me through torturous exercises, and left me exhausted and writhing in agony. Again I waited, shivered, and listened for footsteps.

A tall man breezed in. With piercing black eyes, he glanced in my direction and then stared out of the immense window letting in pale cheerless winter sunshine.

"Well, Mrs. Anderson, we will do a fusion," he said to the window. "Such an operation will relieve your pain, but you will not be able to be active again. Swimming, tennis, horseback riding — any strenuous exercise might break the fusion — which could not be repaired by a repeat operation."

I gasped, "Fusion! What's that?" The surgeon did not answer. "And I'll never be active again. Oh no! This can't be." My tears blotted out the man, the window, and the room.

"We will use a piece of the big bone in your leg," I vaguely heard him say. "We will graft that bone onto your spine in such a way that the vertebra will be apart and supported. This support will release the pressure on the nerve which is causing you pain. But, of course—ah—any quick twist might—ah—snap the fusion. You will—ah—have to be very careful."

"When—will you do this—this fusion?"

"After your final consultation and—ah—arrangements are made at the hospital." He turned back to his desk, picked up his glasses, and left.

A gentle nurse helped me into my clothes. Outside the cubicle Howard met me, gathered me in his arms, and knew without asking what the decision was.

After hours of dazed, agonized, and naturally self-absorbed thinking, the time for our last consultation at the clinic arrived. Howard's warm hand held mine tightly. I was thinking, "No swimming in the cool waters of Shell Lake—no tennis with Howard on summer evenings—no games with our boys—no hikes through the timber in the fall—no walks down snow-covered skidding trails—nothing." A mild-appearing man in a gray-blue suit sat at his desk and scanned reports. In his low gentle voice he asked many questions. Howard answered them while I thought, "Pointless questions. Doesn't he know they've decided to operate?"

Then he spoke directly to me. "My idea is that before a fusion is done you might like to try heat, traction, and massage. Later, if those help, I'd suggest your wearing a brace for a while. The injured vertebrae might produce their own separating structure and relieve the pain."

Howard squeezed my hand so hard I thought he had crushed some bone.

"Yes, of course we'll try anything—anything."

"In six months you may find, however, no relief. An operation would be my final suggestion if all else fails."

Thus began the gloomy year of 1946!

While Sophie, our Finnish housekeeper, took care of the children, I stayed in bed and dangled sandbags from my ankles. Every night, after strenuous tramping in the woods and work at camp, Howard gently massaged my twisted muscles for hours. Slowly the pain became less severe. Late in the summer I walked again, straight as a ramrod and supported by a brace of straps and buckles. But I walked painlessly. And today—without an operation—I am my old self again.

This gloomy year took another shot at us. No sooner had we recovered from the shock of my forced inactivity than we were shaken by another event.

One scorching hot day the children played in the back yard under the lilac tree. Shackled in my brace, I sat in the shade of the house out of the excessive heat and watched them. Dark skies a few days before had promised rains that never fell. The air remained dry and burning hot. The sun still blazed away in the bright sky. Hot winds bent the hollyhocks and scattered fuzz from late-blooming dandelions. Then, with the breeze that raced through the top branches of the tall cedar and swooped low through the asparagus fern, there came the scent of wood smoke.

Wood smoke! *Forest fire!* When the meaning of that smell broke through my senses, dulled by the heat, I began impatiently looking for Howard. He should be coming home from camp. The smelly wood smoke began to settle. Thickly it came down onto everything. Into the house it seeped, entering through open doors and windows.

At length Howard got back from camp, worried and excited. "Fire's broke out. It's hit the timber west and north of camp. Hasn't got to our section yet, but if this wind doesn't stop—if it doesn't rain—I'm afraid the blaze'll travel fast to our stuff. There's all those small pieces piled there that the shackers have been cuttin' all summer. Why, that'd go up like nothing if fire ever started in those slashin's and bark—where they've been peelin' ties and makin' laggin'.—Can I have something to eat right away, Julie? I want to get back out to camp and check the bulldozer—get it greased up and filled with gas. In case we have to use it, it'll be—"

A blast from the fire siren interrupted his words. All tense, we listened to the signal calling for volunteers.

"It must be a real bad blaze, Julie."

Two neighbor boys of high-school age rushed into the kitchen. "Where's the fire, Howard? Can we fight fire?"

"If this wind keeps up we'll need lots of help. Put on caps and heavy shoes, boys.—Long-sleeved shirts, too, if you report for fire fighting."

The phone jangled. Howard grabbed it up. —"Yep, I'll have it ready in an hour. Where do ya want it?—Yep, got a driver, too.—O. K.—I'll pick him up on my way out. We'll cut cross-country with the tractor. It'll be quicker."

"Julie, I've gotta get my bulldozer out to the fire tonight. It's got so bad the Conservation Department's rounding up men and equipment to fight it. Make me a thermos of coffee, will ya?"

Our boys came running in from the sand pile chanting, "Tommy says it's a fire at our daddy's camp. Is it, Mother? Is it, Mother?"

"Not quite, boys," Howard called over his shoulder, "but awful close to it. I'll be home when I get here. But don't worry, Julie. I'm gonna pick Hayward up at camp to drive the tractor."

"Please don't take any chances," were the only words I could get in during the excitement. The car door slammed and Howard was gone, forgetting that he had asked for something to eat and a thermos of coffee. He was anxious only to get started. This fire might consume all we possessed.

Fully dressed and keyed to meet any emergency, I stayed on the davenport that night. Early in the morning I fell asleep. I woke up with burning face and dry throat. A dense pall of smoke hung heavily in the hot air. The dark-gray haze softened the sun's rays, but the heat continued to be almost unbearable. The hours dragged by. No one came home. The fire siren blared its call again for more volunteer fire fighters. The children played outside and took their naps. Still neither Howard nor my brother Hayward came home. Neighbors gathered in our yard to scan the hazy sky to the north and west. The suffocating heat! We exclaimed about it, and speculated about the progress of the fire fighters.

"How do you suppose this got to be such a big fire? How could a fire start 'way off where there aren't any roads?—It's near your camp, isn't it?" I was asked.

"Yes, it's very near our camp—too near to be safe. Lightning maybe—or a fisherman's fire probably started it. No one will ever know—"

At that moment Howard rolled rapidly into the driveway. The motor stopped, and he just sat there slumped and tired. We gathered around the car for news. Deep lines of fatigue showed on Howard's face. He wiped sweat from his red face with a grimy hand. A smudge of black crossed his chin. From one cheek trickled a thick slow ooze of blood. His eyes bloodshot, his clothes reeking of acrid wood-smoke smell, he personified exhaustion.

"Fire under control yet, Howard?"

"Nope. Not yet." Slowly and tiredly he spoke. He slumped lower in the seat. "It's spreadin' faster than they can put it out. They're fightin' with all the men and tractors they can get, and still that blaze keeps comin'. Flames leapin' up all around. Along the ground the fire's burnin' in the dead logs and that powder-dry leaf mold and broken limbs and branches. This wind's carryin' sparks into trees way beyond the fire-fightin' line. If the wind'd only die down," he sighed. I handed him a glass of water and washcloth to wipe his face. "Somebody's gonna get trapped if the wind don't quit. Dead trees just like torches with their tops ablazin'. Power saws are sure comin' in handy now. I picked out a couple o' rampikes to come down and the sawyers brought 'em down in a hurry with those babies. Wish we had more! They're just what we need to stop this roaring ol' scorcher."

"How close to our timber, Howard?"

"It's burnin' pretty close, now. By tomorrow, if this wind continues, I'd say the fire'll be burning Arvid's laggin' piles."

"Where's Arvid and the other shacker?"

"Arvid came in to camp to stay. The others went out to help cut down trees. I gotta get back, too. Just came to tell you that Hayward rested a couple hours at camp and went back out again. The bulldozer and power-saw men are really havin' it tough in this fire. Water's being pumped in from Lake Superior, but it won't do much good till the wind goes down."

"When are *you* going to get some rest, Howard?"

"Not now. Every man is needed, especially guys that know that country. There'll be plenty time later to sleep—when the timber's gone."

There was no need to hold him back by urging him to wash up, eat, and rest. I knew he would not, no matter how I begged. My logger left again in haste.

After the children had their supper and were tucked into bed, I sat silent and alone on the front steps. Each breath of smoke-filled air reminded me that by tomorrow night fire would have flamed through our lagging piles, our decks of hardwood, and our standing timber to leave us ashes—burned-over timber only partly salvageable. Fire! Fire! The thief that steals, that never repays, that can only be foiled by precautions —or days and days and days of dangerous fighting! I sat in the evening darkness numb and weary. I only hoped nothing would happen to Howard and Hayward or the other men. Trees—there were more trees. But fellows like Howard? like Hayward?—Hours went by.

Then I realized the wind had ceased blowing hot on my face. No wind! No wind meant there was a chance—just a slim chance—that the men could gain control of the fire *before* it was too late to save our stuff.

I untangled myself from my cramped position, got up from the steps, and went to bed. The next day found me short of patience and tense with worry and the anxiety of waiting. The dry leaves hung motionless on the trees all day. The rasping song of the cicadas irritated me and frayed my disposition more. I scolded the boys for no reason at all. Nothing went right. The sky, already dark with smoke, became darker. Toward evening the robins began chirping their rain song as they hopped over the dry brown lawn looking for worms. A long deep-throated rumble sounded from the gray skies. A few drops of rain splashed against the kitchen window. Faster and faster the rain drops fell. "Maybe the rain has come in time to save our timber. Oh, I do hope so!"

The next morning after I had washed the breakfast dishes, Howard and Hayward came home. In the basement they shed clothes stained and torn, sweaty and smoky. They wearily climbed the stairs and fell into bed. Half-asleep, Howard mumbled, "Our timber's safe. Don't worry any more."

After sleep had eased their aches, and while coffee bubbled in the coffee maker, I broiled three pounds of steak.—Hayward told us how the bulldozer nearly burned up. "I was a pushin' down trees and turnin' over sod, watchin' all the time to see where the wind was takin' the fire. The old cat purred away good. Just as I was about to push over an old rampike she sputtered a couple-a times and stopped dead. I noticed a blaze in a scraggly old birch to my left and darn close by. I decided to get the heck outta there—darn fast, too.

"The tractor wouldn't budge. She'd run outta gas. I legged it back to where the gas can was, and I never ran so fast in all my life. But it seemed hours before I found the gas can and started back. A fellow with a squirt can came back with me. While I'd been gone that birch, all aflame, had fallen across the tractor. The cushion of the seat was burning and the stink of burning horsehair was worse than all the wood-smoke smell. Sparks were flying through the air. Flames blazed all over the ground.—And me carryin' a can of gas! And there sat a bulldozer worth a couple-a thousand dollars that I couldn't get to!

"The gas tank on the tractor I could see was in the clear. I made a run for it. And made it! The fellow with the squirt can followed me. We covered the birch with water and rolled it off the tractor. He put out the fire on the cushion. While I poured in the gas, I hoped none of them darn sparks'd decide to give me trouble. The motor started right up and we got out of there darn fast. Just as I was backing out, a tree aflame crashed down across the birch, and a cedar caught fire in front of me. Ten minutes later that was sure a hot spot. Boy, oh boy, was *I* lucky!"

"Yah, you sure were. I'm lucky you brought out the tractor. Down where I was—west of you—a lumberjack wasn't that lucky. A widow-maker from a burning tree caught him on the neck and left shoulder. Conservation men took him to the hospital with third-degree burns and a possible fractured skull. The last I heard, he wasn't expected to pull through. He got hurt just after it started to rain."

"Did the fire burn *much* of our timber?" I asked.

"Not at all!—Just barely missed it, though. The fire's not all out yet, but it's surrounded by fire lanes of burned-over stuff. This is one blaze people won't be forgettin' in a hurry, believe me."

For some unaccountable reason our timber stood unscathed by flames that burned many a section elsewhere before they could be controlled. After thinking of our trees as ashes, we suddenly saw them as great wealth again. Happy? We were happier now than ever at any time before the fire.

Not long after the fire we experienced a further sad event, which completed the cycle of *three* incidents for the superstitious lumberjacks.

—Howard had begun winter logging again when tall, lanky, good-hearted Laggin' hit the States in November, 1946. He had gone all the way to Germany up from Africa with the tank division. Laggin' stayed home in Ironwood three days. Then he grabbed his lumberjack shirt and headed for the timber. When he found Howard was still logging, he hit the timber to bunk at Camp Seven. But Laggin' was not to bunk there for long.

The end of his trail came sudden as a shot from a German sniper. The last time I talked to Laggin' was several days after a big snowstorm. I went with Howard to camp just because the day reminded me of other days I had spent in the woods. The roads had not yet been plowed, and we walked through deep snow after we left the car at the highway.

I shall long remember that winter day. Far to the east and north through the hardwood trees we saw the sun shine on the hills at the foot of the Porcupine Mountains. Those patches of sunlight were bright against the otherwise gray and purple hills. Crowns of distant tall pines showed black against the pink-tinted sky. Around the last curve of the road stood the camp buildings, weathered and black. Thin curls of blue smoke twisted from the bunkhouse and cookshanty roofs. The door to the blacksmith shop creaked open and shut, blown by the wind. While Howard gave directions for plowing the roads, I waited in the dusty office.

When we were ready to leave, Laggin' swung out of the bunkhouse with his long stride. We talked for a while until Axel drove up with the bay team hitched to the old dinnering-out dray. "Axel's going to drive us back to the car. It's too long a hike for this late in the afternoon," Howard called.

We climbed into the dray as Laggin' selected a shovel from several propped against the bunkhouse. He started down the road ahead of us to dig out the snowplow. As we passed Laggin' I called out, "Come on and take a sleigh ride out to the car with us, Laggin'."

His big grin spread over his thin brown face. In his slow Finnish way, he called back, "Thanks. Guess I'll just stay here." Those were Laggin's last words to me.

The next day, a November day flooded with bright winter sunshine, Howard left early in the morning for camp in high spirits. He came back home heartsick.

"What's the matter Howard? You look so queer. You're pale as a ghost and—Why you're shaking. What's the matter?"

"It's Laggin'. The kid got killed. I can't believe it. In a matter of—just minutes—he died!"

"Howard!—But—what happened? *Laggin'?*

"An old dead tree broke his neck. Laggin' went out with a power-saw gang. I was there, too. And Axel and Cowboy. They felled a tree that knocked down a dead one alongside of it. I saw the dead tree comin' down. I yelled. Laggin' didn't make it. The trunk caught him right across the back of his neck. I don't know how I got the strength, but I took one step and, Julie, I lifted that tree off Laggin'. We felt his pulse, felt his heart. He didn't breathe. He was dead that quick, Julie. Laggin' must have died the very second the tree hit him."

I couldn't talk. I couldn't believe that Laggin', who had been faithful to all his responsibilities, deserved so tragic an end. Any one of four men there could have been killed. Fate chose Laggin'.

We felt stunned by Laggin's death for months. Yet from the Finnish burial service for him I remembered these beautiful minor-sounding words: "*Hiljaa huojuu korven honka, mielen myrsky nukkuu.*" We both knew them to mean this: "*The pines in the forest are quietly swaying, the storm in his heart is sleeping.*"

Around Stumps the Wild Vines Creep

As our profits from years of good logging increased, so did our desires to own a home of our own. With the same determination that Howard tackles a difficult logging job, he tackled the job of building our home. He incorporated all the latest modern conveniences of heating and general comfort into a Cape Cod home of traditional style. A beautiful satin-smooth pine board from the Big Pine forms our nine-foot fireplace mantel. The pine I helped load out of the mud at the far side of Tula Camp now is wainscoating and woodwork. Our breezeway is enclosed and paneled with veneer from Penokee's mill. It serves the same purpose that the offices at our camps did—a place for power-saw men, sawyers, tractor drivers, cruisers, veneer and pulpwood buyers, and Employer's Mutual Insurance men to shoot the breeze there. The tall tales they tell today concern men known by different names than the men of yesterday, but the length and breadth of these tales are much the same as of those told at Cedar Camp, Tula Camp, and Camp Seven. One change, typical, to be sure, is the replacement of the barrel stove heat by radiant heat from ceiling installation.

Even with the modern conveniences of a radiant heating system, several bathrooms, a recreation room in the basement, a garbage disposal unit in my kitchen sink, and a convenient first-floor bedroom for Howard and me, our home displays much of yesterdays. My favorite spot is in the sun that slants through the French doors which open onto the terrace shaded by our transplanted pines. From there I can enjoy on dull gray days the warmth from our open fire, where my copper teakettle from camp hangs and puffs out whisps of steam. From there, also, I can view our three Currier and Ives prints that hang above the long mantel, or I can gaze out on the natural beauty of our snow-covered pines and recall the majestic beauty of our Big Pine and its haunting minor song.

Books we have read and liked lean in the bookcase against albums of recorded music we enjoy. Two Boston rockers, cozily drawn up by the table which holds my grandmother's old-fashioned kerosene lamp, are fast becoming more antiqued from scratches made by the rivets on cowboy jeans. Cheery sunlight, gradually softening the gray-green, rose, and red colors of the cottage curtains and our plain wine-red upholstered furniture, also is toning down the colors in the hooked rugs I made during long winter days at Tula Camp.

Morning sunlight greets us in the dining room, where the antique wine glasses I have collected on our vacation trips sparkle on their shelves in my Williamsburg-blue corner cabinet. During the daytime little Hayward and yipping Waggy, the children's faithful pal, race in and out from the sandbox while Chris and Hans are at school. In the evenings after the boys are kissed good-night and tucked into bed, Howard and I relax before our fire, burning low, and recall how things have changed since the years we logged and lived and loved at Tula Camp.

Now as always we are up early. Our environment conceals the fact that I am still a logger's wife. But the jangle of the telephone jars me back to reality. And when I put the telephone receiver to my ear, I likely as not will catch the sound of juke-box music and loud laughing. Through this jumble someone shouts, "Is this old man Anderson's place?"

"Yes, but he isn't home now. I expect him around five-thirty."

"Well, I got some pay comin'. Can I get it?"

"Certainly," I say, while Waggy, the mischiefmaker, pulls the tablecloth and a bowl of flowers onto the floor.

"You know, Missus, your old man fired me. He don't know a good jack when he sees one."

"I'm sure you can get your check when he comes home." I think to myself, "Howard probably knows jacks better than this fellow does, but there are only a few around now. Most of the old-timers have hit the trail to the tall timber on the West Coast or to virgin stands they have dreamed about and no man has seen." I hang up the receiver and rush to rescue what is left of the flowers.

–Surely enough, this logger's wife still carries on. Within ten minutes, a rattling car turns into the driveway, at once reminding me of a hot summer night at the Cedar Camp. A jack swings unsteadily out of the front seat. He staggers to the door. I meet him. "Howard is *not* home yet. Please," I beg, "come back later."

Without a word, he strides jauntily down the walk, pats Chris and Hans on their heads, and passes out big silver coins to them and all their playmates. He climbs back into the car and drives away.

In no time at all he telephones again, talks with a thick tongue, and says, "Shay, can I get my dough?"

"Yes." The impatience I feel creeps into the tone of my voice. "But my husband just isn't home yet."

"When in hell does your old man come home?"

I would like to talk back to him in his own language, but I know to do that would be an invitation for a barrage of insults only a *drunken* lumberjack could think of, so I hold my tongue.

Without using any logic, the jack decides to come back and get his check anyway. Without warning, I find him in my kitchen. His sudden presence startles me, but I have learned that drunken lumberjacks are comparatively harmless and easier to handle than stubborn schoolboys. I urge him to wait in his car.

By six o'clock the house smells of beef roast and raspberry pie. By six-thirty the boys have eaten all the rolls and olives from the table. The jack has gone, perhaps to get another drink. At seven Howard comes home, tired, dirty, and all in a slump. I sense that discouraging things have happened. What I imagine is not bad enough, however. My logger and his life of trial!

Hoarsely Howard blurts out the events of the day. "First, the cable on the jammer snapped before the first load was loaded. Then the bay team ran away and broke all their riggin'. A load of logs rolled and tore the stakes all to heck. A gang of gypos buttoned 'er up. The light plant wouldn't start. And, of all things, Axel wants *another cat*."

"What, nobody hurt or killed?" Grim as it sounds, my question reminds him that incidents in the woods are never so bad but that things much worse could have happened.

It is needless to say that on a day like this for Howard the roast turns out to be tough, he wants his pie saved for his lunch tomorrow, and he hears me begin to mumble about the virtues of having a husband who works only eight hours a day at jobs planned for him. And as for me, instead of listening to my favorite concertos, I have to live on through the noise of boys, standing out against Howard's one-sided telephone conversations about the next day's work. And always there is the yip-yipping of Waggy. At the end of days like this I forget that only a few days ago Howard came home at *four* feeling great—announcing with pride and enthusiasm that, "Twelve loads of logs went out of the woods. The skid-ways are full. Veneer has gone up two dollars a thousand feet."

In this latter-day manner, days quickly become weeks; and in good time a season's logging is finished. With favorable weather, few breakdowns, and a good market for logs, the bank account swells. But not, of course, unless Howard has gone out to the woods in the darkness of early morning and has worked all day until dark. Now, as always, the evidences of profits show up only at the end of the summer or winter logging seasons.

If a day comes when Howard says, "Can you be ready to drive to Seattle—or Milwaukee—or Lansing?" then I know that the season's logging has brought profits. A much deserved vacation is in prospect. We leave the children with a dependable housekeeper. Howard and I start out to drive through the Bad Lands, the Black Hills, into Utah, and along the Columbia River to the West Coast. Or we drive to the Straits of Mackinac and cross on the ferry to drive through the Lower Peninsula and to Lansing. Or we go south to Tomahawk, stop at Jim's Landing for one of his stupendous meals, and drive on to Milwaukee.

These new times, I buy ridiculous hats in overdecorated salesrooms, or stop at antique shops to find, say, another wine glass for my collection. We eat lobster, crab, and shrimp until the sauce and butter drip from our ears. Or we fill up on steaks four inches thick. We spend days watching logging as it is done in the big big timber on the West Coast. Or we spend hours searching for just the right toys to take home to our boys. However, on recent vacation adventures such as these we find that we are not real city people or Westerners—just north woods folks. We are happy to come back to our peninsula again to buy and repair equipment and look new timber, and to repair boys' cuts and scratches and bake raspberry pies.

The way of a logger's life with all its vicissitudes still holds Howard and me to this land of iron and of tall timber. I enjoy living in town in our own home with neighbors, grocery stores, schools, and churches nearby. Yet I often yearn to hear the cheery, wild bird calls on bright winter days or the thrilling sound of mating partridge drumming in the spring. I long to hear the ring of the blacksmith's hammer as it used to hit the anvil or the bull cook's sonorous ditty as he hauled the ashes out and returned with chunks of wood. I would like to hear again the drone of the tractor accompanied by the jingling of the skidding horses' rigging when the men came back to Tula Camp at supper time. I remember the cathedral coolness of the tall timber, and I miss the sharp smell of pine mingled with the pungent odor of cedar and the woodsy smell of cut birch and maple. But most of all I miss the "talk" and geniality of the evenings we spent around the barrel stove. Ah! The years we logged and lived at camp —those are years to remember!

The process of gradual change in Upper Peninsula logging camps and logging methods which had been going on for years became greatly accelerated after 1946. These changes are glaringly apparent in our tall timber at the foot of the Kaug Waujoo. There are no jacks in our bunkhouse. Cobwebs hang between the bunks. Dust lies thick on the windowsills, the barrel stoves, and the worn plank floor. Axel has moved into the cookshanty where he lives alone with only aging Tex for company. Axel now takes care of the horses and swamps for the tractor driver. Tex no longer follows the sawyers to the woods. The noise of the power saws is too loud for him, and he is too feeble to shake the quivering branches. The men who drive the tractors and trucks, operate the power saws, or work in gypo gangs sawing and skidding, drive from their homes and carry their lunches with them. A few "batch it" in the shacker's cabins. Old crippled Arvid cuts just enough stuff to pay for his groceries.

The cook's triangle hangs mute by the cookshanty door. There are no husky jacks to answer the call to supper. There are no more tall tales told or poker games played in the warmth of the smelly bunkhouse. The barrel stoves do not glow cherry-red in winter with the heat of burning hardwood chunks. Our office stands empty except for rusty crosscut saws, and extra parts and sharpening tools for the power saws. Six stalls in the barn are empty. A new team whose names I do not know occupy two stalls. A rusty forge, deserted years ago by Paul, stands in its usual corner in the gloomy blacksmith shop. The pony harness, rotted from its hook, lies on the floor. Rigging that needs repairs is piled by the doorway.

The "fireproof" Burton House, the landmark of boom logging days in this region of the Upper Peninsula (even though it stood across the Montreal River in Wisconsin) lies entombed in ashes. Part of its chimney still stands, marking the grave of faded splendor, recent vice, and bare lumberjack quarters. And yet, for all these changes, the logs still roll out of the Upper Peninsula woods and down the highways to the mills. Such is the great natural wealth of our inexhaustible logging region.

Felling trees and bucking limbs with power saws instead of crosscuts, using magnesium wedges instead of hardwood wedges, loading logs with power jammers instead of the pull of crosshaul teams, skidding more logs with tractors than with skidding teams, and hauling logs with more and larger trucks, Howard cuts deeper into the timber every year. At one high place on a woods road beyond Section Thirteen, where Howard logged this summer, the blue of Lake Superior can be seen when the maple buds have not yet burst out in lovely greens. Along both sides of his roads more than fifteen skidways are filled with logs—birch, maple, basswood, hemlock, and some pine. At a sharp curve by the end of a skidding trail gouged into the moist earth lies the largest ash log I have ever seen. Equipment stands ready to be used on Monday. Jammers are tipped back against waiting decks. The hoist, mounted on a truck and used for loading instead of teams or tractors, is parked at an empty skidway. Cant hooks and rigging adorn the logs. Forgotten on a stump, two cups from the cookshanty tell of lunches in the woods. The tops of tall white pines to which, as yet, there are no roads can be seen against the blue north sky. Preparations for winter logging are apparent in freshly graded roads. The drama and excitement of logging still exist.

To the south and east lies our old Tula Camp, flooded

with sunshine. Long ago everything there was cut and hauled —ties, posts, lagging, and logs. The tarpapered buildings are torn down. Tula Camp is now only a clearing in the cut-over timber. Giant mullein stalks tower above the lavender thistle blossoms. Lost rigging is now forever only a part of this ground covered over by daisies, milkweeds, and touch-me-nots. Around the stumps creep the wild morning glory vines. On the shady side of these stumps sweet red raspberries grow. Goldenrod and tall fire grass have grown in the middle and sides of the woods roads and have gradually crept in to cover the deep ruts.

The planks across the "soft spot" gleam like strips of silver, weathered by the rain, the sun, the snow. Brush piles lie disguised by grass and vines to trap occasional berry pickers. The calls of the finch and grosbeak interrupt the hammering of the woodpeckers as before. Eerie sounds of leaning rampikes rubbing on the scraggly swaying trees that support them haunt these woods.

Around the Stumps
the Wild Vines Creep